Chronobiology and Obesity

Marta Garaulet • Jose M. Ordovás
Editors

Chronobiology and Obesity

Springer

Editors
Marta Garaulet
Department of Physiology
University of Murcia
Murcia, Spain

Jose M. Ordovás
Tufts University School of Medicine
Boston, MA, USA

ISBN 978-1-4614-5081-8 ISBN 978-1-4614-5082-5 (eBook)
DOI 10.1007/978-1-4614-5082-5
Springer New York Heidelberg Dordrecht London

Library of Congress Control Number: 2012950037

Printed on acid-free paper

Springer is part of Springer Science+Business Media (www.springer.com)

A Historical Perspective

The skeletal remains of *Ardipithecus ramidus*, an extremely ancient hominid and the ancestor of present day humans, were discovered in the Afar Depression of Ethiopia in 1992. This bipedal hominin from the early Pliocene period lived an estimated 4.4 million years ago and survived in a photoperiodic environment similar to that of the present day, i.e., with a 24-h light:dark cycle. If she (the bones of the original *A. ramidus* uncovered were from a female) did not already possess at least a primitive biological clock in her brain, which by today's standards was small, she and subsequent generations had the intervening 4.4 million years to evolve and perfect the function of this important set of neurons. The choice of the light:dark cycle on which to rely for regulation of the clock was clearly appropriate given that other 24-h rhythms, e.g., the ambient daily temperature cycle, are much less reliable. Today, circadian rhythms, dictated by the central oscillator in the brain, are known to be key features of the physiology of all vertebrates.

While 24-h rhythms in human physiology have been prevalent for eons, the actual location of the circadian timer in the brain was only recently precisely defined. Without having a clue as to the function of a small group of neurons located bilaterally near the midline above the optic chiasm at the diencephalic-telencephalic junction, EA Speigel and H Zweig in 1917 named this nuclear group the suprachiasmatic nucleus (SCN). In the first 40 years after its discovery, other neurobiologists also identified this nucleus and assigned it different names, none of which stuck. In the early 1970s, RY Moore and colleagues noted that axons originating in the retinas coursed with the optic nerve and terminated in the SCN; this set of specialized axons is appropriately named the retinohypothalamic tract (RHT). During the decade of the 1970s, it was also noted by a number of scientists that destruction of the suprachiasmatic nucleus was associated with abolition of a number of 24-h rhythms, e.g., drinking, locomotion, and adrenocorticosterone. Thus, the SCN was identified as the central generator of 24-h rhythms. In the last 15 years, research in this system has proceeded at a feverish pace with new knowledge of the molecular mechanisms of the clock and how it communicates with peripheral timing processes, which likely exist in all cells throughout an organism, accumulating almost weekly.

The discovery of the perikarya that gives rise to the axons which specifically form the RHT was forthcoming only within the last 15 years. The axons originate from a novel set of retinal ganglion cells that constitute less than 1 % of the total retinal ganglion cell population. These highly specialized cells are referred to as intrinsically photoreceptive retinal ganglion cells (*ip*RGC). The *ip*RGC contain their own unique photopigment, melanopsin, and they subserve what has come to be known as circadian vision. This system is functionally distinct, for the most part, from the photoreceptive rods and cones which function in visual vision. Like the regular retinal ganglion cells, of which there are possibly a dozen subtypes, *ip*RGC also exhibit at least anatomical diversity with several subtypes being categorized.

The photopigment, melanopsin, which functions in phototransduction in the *ip*RGC, exhibits a blue light spectral sensitivity (~480 nm). Multiple isoforms of melanopsin have been discovered in some species. Functions subserved by the *ip*RGC via the SCN and other nuclear groups in the central nervous system include photoentrainment of the SCN, pupillary light reflexes, light regulation of sleep processes and suppression of melatonin production.

The molecular processes at the level of the central circadian pacemaker, the SCN, are also being unraveled. The clock processes in these cells are autonomous and maintain their own approximately 24-h rhythm. This rhythm persists in the absence of environmental stimuli thereby always providing some degree of temporal organization. The major factor that ensures regular 24-h cycles is that the clock can be light entrained, i.e., "set" to local time; moreover, the clock exhibits temperature compensation, i.e., it is not influenced by fluctuations of ambient or internal temperatures.

The importance of clock processes is emphasized by their remarkable ubiquity. Clocks exist in all organisms ranging from prokaryotic microbes, to plants, and up to humans. In multicellular organisms, peripheral cells also contain clocks that are subservient to circadian messages from the SCN.

The evidence is now indisputable that regular circadian changes in physiology, behavior, and metabolism are absolutely essential for the optimal function of every organ and organism. Disregulation of these basic rhythms, i.e., circadian disruption, circadian desynchrony, chronodisruption, etc., obviously negatively impact the health of individuals as reviewed in the current volume. Unfortunately, the single most important environmental factor that drives the circadian clock via the previously described *ip*RGC and the SCN, i.e., the light:dark cycle, is becoming badly corrupted due to the widespread use of artificial light. There is very likely a pathophysiological "price to pay" for this corruption as illustrated in the chapters in this book.

The likelihood of this bad situation resolving itself in the near future seems remote. The use of artificial light is becoming progressively more commonplace and the end of darkness may be near. How humans will solve the problem of the physiological disturbances in these basic and intrinsic rhythms and avoid the pathological consequences is not yet apparent. Even if we successfully deal with this problem on Earth, when we take up residence on other planets where the light:dark environment will be very different from that on Earth, we again will be met with this continuing challenge.

San Antonio, TX, USA Russel J. Reiter

Preface

Chronobiology, defined as the science that studies the *circadian* (around a day) rhythms of biological beings, is a relatively new science which was first discovered in the eighteenth century. It came out with Linnaeus, who designed a beautiful "floral clock" that represented the hours of the day depending on the time that flowers open their petals.

Among human beings, the circadian rhythms are so inherent to our lives that we do not even notice them. Perhaps this is the reason why in the medical practice the circadian variability of the hormones, metabolites, and physiological behaviors or the relevance of time in the presence or absence of different pathologies has been practically ignored in the past.

However, two important circumstances have allowed us to see a revolutionary change in this trend. The first is the discovery that in our bodies, apart from the central clock, most of the organs and systems have their own clocks (peripheral clocks) that can work without the influence of the suprachiasmatic nucleus (SCN).

The second essential circumstance is that over the past two decades, biochemical, genetic, and molecular studies have been making substantial advances towards the elucidation of the molecular basis of rhythmicity in living things.

Riding on the wave generated by the seminal studies in the 1970s focusing on the circadian variability of hormones such as cortisol, melatonin, or growth hormone (GH), or those related to the discovery and description of the physiological bases of the SCN, current chronobiology has dramatically evolved thanks to the new genetic and molecular biology techniques.

Now we are able to study the expression of the known clock genes implicated in the circadian machinery. We already know that, in mammals, the core components of the clock molecular machinery operate in almost all cells of the body through a complex network of transcription–translation loops and modulate the expression of specific target genes and their products to oscillate in 24-h rhythm.

Moreover, experimental models are allowing us to assess clock genes expression not only in the living animal but also outside of the body (in vitro techniques), and we are also able to analyze the 24-h fluctuations in gene expression and to assess the presence or absence of a peripheral clock in the different organs and tissues (see Chap. 2).

From the genetic epidemiology point of view, the study of single nucleotide polymorphisms (SNPs) is contributing to the identification of the genetic background of chronotypes (morningness or eveningness), sleep alterations, or seasonal mood disorders. All these advances have allowed researchers to find the *relevant link which exists between chronobiology and obesity, which is the main goal of the current book.*

While reading the different chapters of this book, we will have the opportunity to read from Professor Russel Reiter how regularly alternating periods of light and darkness, such as that which normally occurs with the rising and setting of the sun, are essential for the maintenance of undisturbed circadian rhythms in all organisms including humans.

As we continue, Professor Alfred W Turek, we will be able to read the beautiful story about the discovery of the Clock mutant and the first mammalian clock gene and the links to obesity: Starting with animal #25. As Professor Turek says in his introductory chapter, "Unexpected at the time, the discovery of the *Clock* mutant animal would eventually lead to an entire new approach to the study and treatment of obesity." "Although still in the early stages of discovery, linking circadian clock genes to energy regulation has clear implications for future studies on body weight regulation at the mechanistic level, as well as for the development of new therapeutic approaches for combating the epidemic of obesity, as well as metabolic disorders, including diabetes."

The next chapter written by Juan A Madrid will give us some basic issues in chronobiology, including an explanation of the complex function of the circadian clock and an introduction of the different techniques more widely used in the current moments to assess circadian rhythmicity in the individuals.

After reading the chapter of Purificación Gómez-Abellán and Marta Garaulet we will know that "Time Is of the Essence" in adipose tissue. Energy metabolism and circadian systems have evolved together over millions of years to optimize internal coordination among multiple physiological and molecular processes. Therefore, further investigations of circadian rhythms in adipose tissues will provide insight into the physiology of energy homeostasis and the etiology of metabolic diseases such as obesity.

In another chapter, JA Madrid and R Reiter among other authors will introduce us the concept of chronodisruption: causes and consequences; and, with the chapter of Oren Froy, we will be aware of how important is the time of eating in obesity. In these same lines, La Fleur SE will introduce us to the relationship between chronobiology and the glucose intolerance, a Metabolic Syndrome risk so tidily connected to obesity. And thanks to Henrik Oster words we will know that sleep is also an important issue in obesity.

Finally, in the last steps of this amazing walk, we will be able to understand the genetic bases which are behind chronobiology and obesity with the explanations of Marta Garaulet and Jose M Ordovás, and those of Silvia Sookoian and Carlos J Pirola, who used systems biology approaches to integrate genomic, molecular, and physiological data to interpret putative circadian rhythmic pathways suspected to play a role in the etiology of the metabolic syndrome.

In summary, we expect that after reading this book the reader will arrive to the same conclusion as we did about the important connection existing between obesity and chronobiology, although as Professor Turek says in his introductory chapter: "Thus, as with an iceberg, whose tip provides only a glimpse (roughly 10 %) of what is below the surface, the importance of circadian temporal organization for health and disease could be profound. Indeed, what we are going to do with time and medicine at the beginning of the twenty-first century is what Einstein did with time and physics at the beginning of the twentieth century. What we have now is just the tip of the iceberg."

We hope that you will enjoy while reading this book, as much as we had enjoyed while writing it. If you would like to read biographies of the chapter authors, please go to http://Extras.Springer.com

Murcia, Spain Marta Garaulet

Contents

Contributors

Oren Froy, PhD Faculty of Agriculture, Food and Environment , Nutrigenomics and Functional Foods Research Center, Institute of Biochemistry, Food Science and Nutrition,The Hebrew University of Jerusalem, Rehovot, Israel

Marta Garaulet, PhD Department of Physiology, University of Murcia, Murcia, Spain

Purificación Gómez-Abellán, PhD Department of Physiology, University of Murcia, Murcia, Spain

Jana Husse, PhD Circadian Rhythms Group, Max Planck Institute for Biophysical Chemistry, Göttingen, Germany

Andries Kalsbeek, PhD Department of Endocrinology and Metabolism, Academic Medical Center, University of Amsterdam, Amsterdam, The Netherlands

Hypothalamic Integration Mechanisms, Netherlands Institute for Neuroscience, Amsterdam, The Netherlands

Susanne E. la Fleur, PhD Department of Endocrinology and Metabolism, Academic Medical Center, University of Amsterdam, Amsterdam, The Netherlands

Xiaoyan Liu Department of Cellular and Structural Biology, University of Texas Health Science Center at San Antonio, San Antonio, TX, USA

Juan Antonio Madrid Pérez, PhD Chronobiology Laboratory, Faculty of Biology, Department of Physiology, University of Murcia, Murcia, Spain

Lucien C. Manchester, PhD Department of Cellular and Structural Biology, University of Texas Health Science Center at San Antonio, San Antonio, TX, USA

Jose M. Ordovás, PhD Jean Mayer, US Department of Agriculture, Human Nutrition Research Center on Aging, Tufts University School of Medicine, Boston, MA, USA

The Department of Epidemiology and Population Genetics, Centro Nacional Investigaciones Cardiovasculares (CNIC), Instituto Madrileño de Estudios Avanzados en Alimentación (IMDEA-FOOD), Madrid, Spain

Henrik Oster, PhD Circadian Rhythms Group, Max Planck Institute for Biophysical Chemistry, Göttingen, Germany

Medical Department I, University of Lübeck, Lübeck, Germany

Carlos J. Pirola, PhD Department of Molecular Genetics and Biology of Complex Diseases, Institute of Medical Research A Lanari-IDIM, University of Buenos Aires-National-Council of Scientific and Technological Research (CONICET), Ciudad Autónoma de Buenos Aires, Argentina

Russel J. Reiter, PhD Department of Cellular and Structural Biology, University of Texas Health Science Center at San Antonio, San Antonio, TX, USA

Sergio A. Rosales-Corral, MD Department of Cellular and Structural Biology, University of Texas Health Science Center at San Antonio, San Antonio, TX, USA

Silvia Sookoian, MD, PhD Instituto de Investigaciones Médicas A. Lanari, Av. Combatiente de Malvinas, Buenos Aires, Argentina,

Department of Clinical and Molecular Hepatology, Institute of Medical Research A Lanari-IDIM, University of Buenos Aires-National Council of Scientific and Technological Research (CONICET), Ciudad Autónoma de Buenos Aires, Argentina

Dun-Xian Tan, MD, PhD Department of Cellular and Structural Biology, University of Texas Health Science Center at San Antonio, San Antonio, TX, USA

Anthony H. Tsang, MSc Circadian Rhythms Group, Max Planck Institute for Biophysical Chemistry, Göttingen, Germany

Fred W. Turek, PhD Department of Neurobiology, Weinberg College of Arts and Sciences, Chicago, IL, USA

Center for Sleep & Circadian Biology, Northwestern University, IL, USA

Chapter 1
Discovery of the *Clock* Mutant and the First Mammalian Clock Gene and the Links to Obesity: Starting with Animal #25

Fred W. Turek

Abstract The discovery of the *Clock* mutant mouse in 1994 and the identification and cloning of the gene underlying the mutation, named *Clock*, was a landmark finding in the history of the field of mammalian circadian rhythms. Unexpected at the time, the discovery of the *Clock* mutant animal would eventually lead to an entire new approach to the study and treatment of obesity. The report in 2005 that the *Clock* mutant animal is obese and shows signs of the metabolic syndrome opened up an entire new field of obesity research. Although still in the early stages of discovery, linking circadian clock genes to energy regulation has clear implications for future studies on body weight regulation at the mechanistic level, as well as for the development of new therapeutic approaches for combatting the epidemic of obesity, as well as metabolic disorders, including diabetes.

The discovery of the *Clock* mutant mouse in 1994 [27] and the identification and cloning of the gene underlying the mutation, named *Clock* [4, 9], was a landmark finding in the history of the field of mammalian circadian rhythms. Unexpected at the time, the discovery of the *Clock* mutant animal would eventually lead to an entire new approach to the study and treatment of obesity. The report in 2005 that the *Clock* mutant animal is obese and shows signs of the metabolic syndrome [25] opened up an entire new field of obesity research. Although still in the early stages of discovery, linking circadian clock genes to energy regulation has clear implications for future studies on body weight regulation at the mechanistic level, as well as for the development of new therapeutic approaches for combatting the epidemic of obesity, as well as metabolic disorders, including diabetes.

F.W. Turek, Ph.D. (✉)
Department of Neurobiology, Weinberg College of Arts and Sciences, Chicago, IL, USA

Center for Sleep & Circadian Biology, Northwestern University, 2205 Tech Dr., Evanston, IL 60208-3520, USA
e-mail: fturek@northwestern.edu

M. Garaulet and J.M. Ordovás (eds.), *Chronobiology and Obesity*
DOI 10.1007/978-1-4614-5082-5_1,

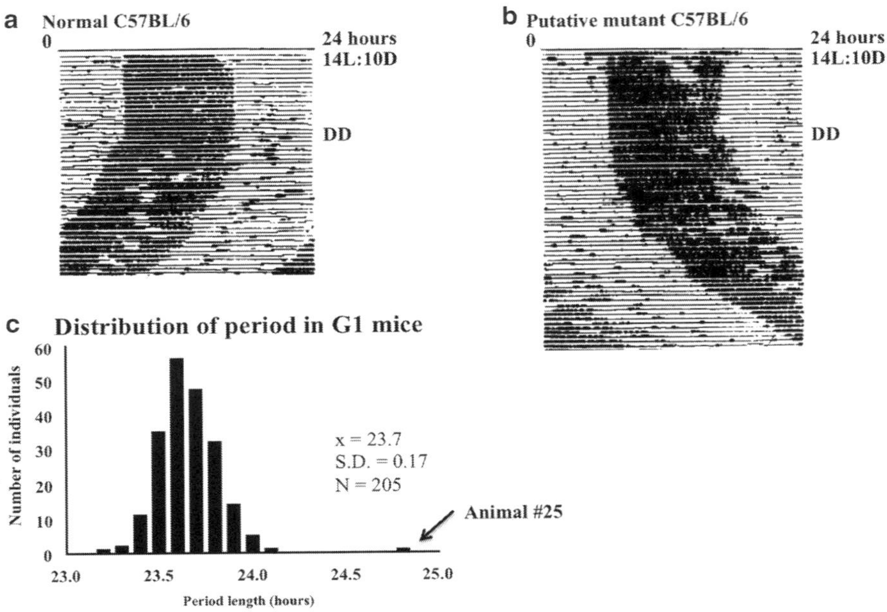

Fig. 1.1 The discovery of the *Clock* mutant mouse: animal #25. (**a**) Activity record plotted over 24 h (*left* to *right*) and over many days (*top* to *bottom*) of a representative wild-type C57BL/6J initially exposed to an LD 14:10 light–dark cycle before being transferred to constant darkness (DD). The free-running period of this mouse was 23.7 h. (**b**) Activity record of animal #25 from the first batch of 42 animals that were the offspring of mutagenized male mice plotted in the same way as in (**a**). The free-running period of this putative mutant was 24.7 h. (**c**) The distribution of the free-running period for the initial 205 offspring of mutagenized male mice (G1 mice) that collectively showed a mean (x) free-running period of 23.7 h with a very tight standard deviation (SD) of only 0.17 h. Animal #25, with a period of 24.7 h, was six standard deviations from the mean. Adapted from [27]

Dove shipped to Northwestern; in fact, it was the 25th mouse she set up to record locomotor activity first in LD then in DD. In DD, wild-type C57BL/6J male mice have a free-running period of ~23.7 h; mouse #25 had a free-running period (Tau) of ~24.7 h, some *six* standard deviations from the mean of WT mice (Fig. 1.1). While we were pleasantly surprised to find such a circadian putant (i.e., putative mutant) mouse so quickly, we held our breath to see if the mutant animal could produce offspring (worthless if he could not) and whether the mutant phenotype would be genetically transmitted. Indeed, the putant was a mutant, as the founder *Clock* mutant mouse (a male) produced 24 offspring from four matings: 13 of the mice had a wild-type circadian phenotype in DD (*Tau* ~23.7 h) while 11 mice had a mutant circadian phenotype (*Tau*=24.7 h). These 11 "heterozygous" mice (the founder mutant could only be carrying a single copy of the mutant gene received from his mutagenized father were a mix of male and females, and thus, two heterozygous mutant mice could be mated to produce offspring. The first litter of such a heterozygous mating produced four offspring that collectively showed the classical Mendelian genetic ratio for a single gene dominant or semidominant mutation from a heterozygous × heterozygous cross (1:2:1), with one littermate having a 23.7-h free-running period in DD (i.e., homozygous wild-type), two mice showing the founder's 24.7-h period (i.e., heterozygous) and one mouse showing a heretofore never seen circadian phenotype (i.e., homozygous mutant) a free-running period of 27–28 h in DD before the activity rhythm often decayed into arrhythmicity. Number 25 was indeed a *Clock* mutant animal.

In just 3 years after the 1994 *Science* publication, the *Clock* mutant gene was identified and cloned using a variety of molecular and genetic techniques and strategies, including positional cloning and gene rescue [4, 9]. While *Clock* was the first mammalian circadian clock gene to be cloned, two laboratories soon published papers demonstrating by sequence homology that mammals (mice and humans) harbored a homologue of the *Drosophila Per* gene (in fact three copies of the *Per* gene) [18, 22]. The race was on and the core members of the TTFL were found in both flies and mammals with many of the genes and proteins being similar in sequence and function in these evolutionary distant species. These breakthrough studies in flies and mice, and others in *Neurospora* and cyanobacteria, led *Science* magazine to call the circadian clock gene story the number one "biomedical" breakthrough of the year in 1998 [17]. The discovery of the mammalian *Clock* and *Per* genes, and homologous genes in flies, led to a domino effect with many of the core circadian clock genes that compromise the TTFL being identified in just a few years.

Surprise: The Molecular Circadian Clock Is Everywhere

Although there were a few hints that other areas of the brain or peripheral tissues/organs might contain the circadian clock machinery in mammals (particularly in the retina; see [23]), the overall working hypothesis was that the SCN cells contained intrinsic circadian rhythm generating capabilities, and at best, downstream tissues and organs might contain damped circadian oscillators that would be driven by the central clock in the SCN (Fig. 1.2a). Thus, it was a surprise to find that the core molecular circadian clock genes (*Clock*, *Per1*, *Per2*, *Per3*, *Bmal1*, *Cry1* and *Cry2*) could be found in many, many different tissues and organs (Fig. 1.2b). Furthermore, the expression of these clock genes was rhythmic, even when the tissue and organs were removed and clock gene expression was followed in vitro for many circadian cycles [19, 21].

The molecular circadian clock was found to be oscillating in many tissues involved in body weight regulation, including brain areas involved in food intake, as well as key peripheral metabolic organs such as the liver, muscles, adipose tissues and pancreas. Furthermore, this clock was not just keeping time for itself; hundreds of "clock-controlled genes" (CCGs) were found to cycle in a tissue-/organ-dependent manner and to be under the regulation of the local molecular clock machinery [3, 13], which in turn is under the direct or indirect rhythmic control of the master circadian clock in the SCN.

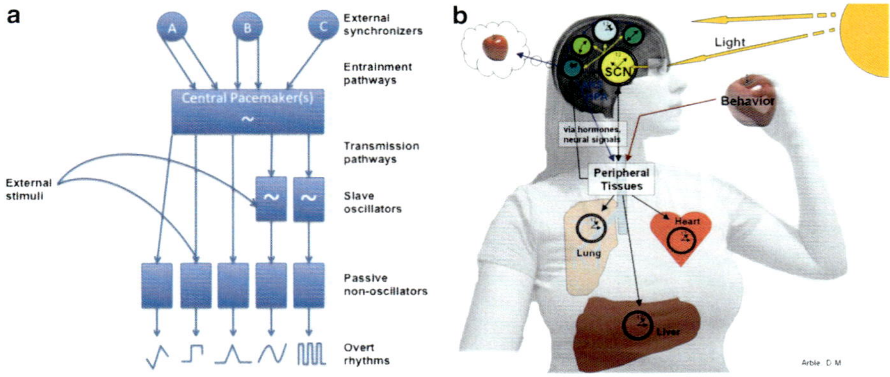

Fig. 1.2 (**a**) Schematic representation of a putative model for the organization of the mammalian circadian system. Before the discovery of the core molecular clock being presented in most, if not all, the tissues and organs of the body, it was speculated, with little experimental support, that there might be circadian clocks outside of the central circadian pacemaker in the SCN. Adapted from [24]. (**b**) A more up-to-date version of the organization of the circadian clock system was revealed when it was discovered that the molecular circadian clock was everywhere and that in vitro, various tissues/organs remain rhythmic. In this more updated mode, light entering the retina stimulates specialized photoreceptors and send signals to the suprachiasmatic nuclei (SCN) via the retinohypothalamic tract. The SCN then orchestrates the timing of other brain regions. These brain regions can then influence one other, cause behavior changes and send timing cues to peripheral tissues using hormones and neural signals through the hypothalamic–pituitary–adrenal (HPA) axis and the autonomic nervous system (ANS). Behaviors such as feeding can also directly influence peripheral tissue timing. Hormones and neural signals originating from the periphery can then feedback to the SCN and other brain regions. If the molecular clock is present in all cells, it is not clear if there are any passive non-oscillatory tissues/organs of the body as was speculated before the molecular clock was discovered to be everywhere. Adapted from [5]

How #25 Led to New Links Between Circadian Clock and Obesity/Metabolic Regulatory Genes

It has been recognized for many years that shift-workers are more likely to suffer from obesity and cardiometabolic disorders, as well as neurological disorders. However, it has been, and is still, difficult to disentangle the negative health effects associated with the circadian disruption that occurs in shift-workers with multiple other factors that can underlie the adverse effects of shift-work on health, such as sleep loss, use of alcohol and drugs, disruption of social and family networks. While a number of animal models have linked the environmental disruption of circadian rhythms to increased mortality and adverse health [14], the ability to take a genetic approach to study such linkages was quite limited until the discovery of the *Clock* mutant animal. However, even with the discovery of the *Clock* and other circadian mutant mice, few early studies took advantage of using animals carrying mutations in core circadian clock genes for unraveling the importance of disrupting the circadian clock on health and disease. This may have been, in part, due to the fact that early studies of the circadian mutant mice focused on using them to unravel the nature of the genetic and molecular circadian clock, and how mutations affect the expression of circadian rhythms under different environmental conditions. Nevertheless, a few studies were carried out to determine how the *Clock* mutation affected the sleep–wake cycle [12] and the regulation of the pre-ovulation LH surge [10, 11]: two rhythms known to be strongly under the control of the circadian clock.

Given the well-known importance of the circadian clock in regulating the timing of food intake, determining the effects of the *Clock* mutation on energy regulation seems like an obvious area of research that should have been pursued soon after the discovery of the *Clock* mutant animal. However, except for some interest in the seasonal control of body weight and hibernation, I had very little interest in studying how the mutation might affect body weight regulation. In fact, I was pushing my students to carry out studies on how the *Clock* mutation might affect aging (or how aging might affect the clock of aging *Clock* mutant mice) when I serendipitously fell into the world of obesity. One of my students, Amy Easton, who was close to obtaining her PhD was carrying out a few pilot studies on aging (Amy was the first to show that the *Clock* mutation had an effect on cognitive function, in this case, increasing exploratory activity and escape-seeking behavior; [8]) when she reported to me that she did not like working with *Clock* mutant mice because they got fat, which interfered with their ability to run on the running wheel. I remember making some glib remark that maybe we should study obesity and probably would have abandoned that idea were it not for the recruitment of Joe Bass to a Northwestern-affiliated hospital who had recently set up his laboratory down the hall from Takahashi and me. When we brought the "*Clock* mutant gains weight" story to Bass, and asked him what we should do, he encouraged us to study body weight gain in mutant animals on a high-fat diet. Thus started our collaboration and the publication of our 2005 paper in *Science*, "Obesity and Metabolic Syndrome in Circadian Clock Mutant Mice."

Quite often in the history of the field of rhythms a circadian clock researcher links the circadian world to a new area of biomedical research, but fails to recruit this broader biomedical world into the circadian world. I believe this happened in the case of our 1997 paper [14] where we showed in a dramatic way that by phase shifting the LD cycle on a weekly basis we induced increased and earlier mortality in cardiomyopathic hamsters; results that did not capture the interests of the broader world of biomedical researchers interested in the heart and heart-related disease. This was not the case with the metabolism establishment. When they learned that the *Clock* mutant animal was obese and developed signs of the metabolic syndrome, many of the top researchers in the field moved part of their research interests into the circadian world. Indeed, soon after the publication of our *Science* paper in January 2006, Bart Staels [20], a well-known nuclear receptor and metabolic researcher, published a News and Views report in *Nature Medicine* with the clever title, "When the *Clock* stops ticking, metabolic syndrome explodes." Something else exploded: there was now an explosion of interest within the obesity and metabolism community to uncover how clock and metabolic genes interact and regulate one another and how these interactions underlie body weight and energy regulation as well as related diseases, including obesity, diabetes, and cardiometabolic disorder.

Much of this interest has focused on daily remodeling of histone proteins [1, 26], as well as nuclear receptors, NAD biosynthesis, and nutrient sensors [5, 6]. Early studies by Ueli Schibler and colleagues were among the first to demonstrate that circadian clock genes interact with metabolic genes [15], laying the foundation for the explosion of research that occurred after the descendants of #25 were found to be obese.

Summary

When mouse #25 was identified as the *Clock* mutant animal in 1994, I don't think any of us realized how circadian researchers would rapidly uncover the core mammalian molecular clock TTFL as well as a host of other interacting core circadian clock loops. Similarly, when *Science* editors selected the circadian clock gene story as the number one biomedical breakthrough of the year in 1998, few could see that just over the horizon were lurking many, many stories of how the core molecular circadian clock genes are interacting with many, many cellular processes underlying a wide variety of disease states. Clearly, our understanding of how circadian clock genes interact with metabolic genes and pathways has been at the forefront of opening up a new area of medicine: circadian medicine. While many studies have uncovered molecular links between circadian and metabolic genes and the importance of transcriptional regulation, fewer studies have attempted to examine such links between the core circadian clock genes and the regulatory genes and gene networks underlying other physiological systems. Thus, as with an iceberg, whose tip provides only a glimpse (roughly 10 %) of what is below the surface, the importance of circadian temporal organization for health and disease could be profound. Indeed, bringing time to medicine at the beginning of the twenty-first century may transform medicine in the way Einstein did in bringing time to physics at the beginning of the twentieth century. What circadian biologists have now is just the tip of the iceberg.

References

1. Alenghat T, Meyers K, Mullican SE, Leitner K, Adeniji-Adele A, Avila J, Bućan M, Ahima RS, Kaestner KH, Lazar MA (2008) Nuclear receptor corepressor and histone deacetylase 3 govern circadian metabolic physiology. Nature 456(7224):997–1000
2. Allada R, Emery P, Takahashi JS, Rosbash M (2001) Stopping time: the genetics of fly and mouse circadian clocks. Annu Rev Neurosci 24:1091–1119
3. Andrews JL, Zhang X, McCarthy JJ, McDearmon EL, Hornberger TA, Russell B, Campbell KS, Arbogast S, Reid MB, Walker JR, Hogenesch JB, Takahashi JS, Esser KA (2010) CLOCK and BMAL1 regulate MyoD and are necessary for maintenance of skeletal muscle phenotype and function. PNAS 107(44):19090–19095
4. Antoch MP, Song EJ, Chang AM, Vitaterna MH, Zhao Y, Wilsbacher LD, Sangoram AM, King DP, Pinto LH, Takahashi JS (1997) Functional identification of the mouse circadian Clock gene by transgenic BAC rescue. Cell 89(4):655–667

5. Arble DM, Ramsey KM, Bass J, Turek FW (2010) Circadian disruption and metabolic disease: findings from animal models. Best Pract Res Clin Endocrinol Metab 24:785–800
6. Bass J, Takahashi JS (2010) Circadian integration of metabolism and energetics. Science 330(6009):1349–1354
7. Dunlap JC (1999) Molecular bases for circadian clocks Cell 96:271–290
8. Easton A, Arbuzova J, Turek FW (2003) The circadian *Clock* mutation increases exploratory activity and escape-seeking behavior Genes Brain Behav 2(1):1 –19
9. King DP, Zhao Y, Sangoram AM, Wilsbacher LD, Tanaka M, Antoch MP, Steeves TDL, Vitaterna MH, Kornhauser JM, Lowrey PL, Turek FW, Takahash JS (1997) Positional cloning of the mouse circadian Clock gene. Cell 89:641–653
10. Miller BH, Olson SL, Turek FW, Levine JE, Horton TH, Takahashi JS (2004) Circadian clock mutation disrupts estrous cyclicity and maintenance of pregnancy. Curr Biol 14(15):1367–1373
11. Miller BH, Olson SL, Levine JE, Turek FW, Horton TH, Takahashi JS (2006) Vasopressin regulation of the proestrous luteinizing hormone surge in wild-type and Clock mutant mice. Biol Reprod 75(5):778–784
12. Naylor E, Bergmann BM, Krauski K, Zee PC, Takahashi JS, Vitaterna MH, Turek FW (2000) The circadian *Clock* mutation alters sleep homeostasis in the mouse. J Neurosci 20(21):8138–8143
13. Panda S, Antoch MP, Miller BH, Su AI, Schook AB, Straume M, Schultz PG, Kay SA, Takahashi JS, Hogenesch JB (2002) Coordinated transcription of key pathways in the mouse by the circadian clock. Cell 109(3):307–320
14. Penev PD, Zee PC, Turek FW (1997) Quantitative analysis of the age-related fragmentation of hamster 24-h activity rhythms. Am J Physiol 273:R2132–R2137
15. Preitner N, Damiola F, Lopez-Molina L, Zakany J, Duboule D, Albrecht U, Schibler U (2002) The orphan nuclear receptor REV-ERBalpha controls circadian transcription within the positive limb of the mammalian circadian oscillator. Cell 110(2):251–260
16. Ralph MR, Menaker M (1988) A mutation of the circadian system in golden hamsters. Science 241(4870):1225–1227
17. Science (1998) Breakthrough of the year. Science 282(5397):2157–2161
18. Shearman LP, Zylka MJ, Weaver DR, Kolakowski LF Jr, Reppert SM (1997) Two period homologs: circadian expression and photic regulation in the suprachiasmatic nucleus. Neuron 19(6):1261–1269
19. Shearman LP, Sriram S, Weaver DR, Maywood ES, Chaves I, Zheng B, Kume K, Lee CC, van der Horst GT, Hastings MH, Reppert SM (2000) Interacting molecular loops in the mammalian circadian clock. Science 288(5468):1013–1019
20. Staels B (2006) When the *Clock* stops ticking, metabolic syndrome explodes. Nat Med 12(1):54–55
21. Stokkan KA, Yamazaki S, Tei H, Sakaki Y, Menaker M (2001) Entrainment of the circadian clock in the liver by feeding. Science 291(5503):490–493
22. Sun ZS, Albrecht U, Zhuchenko O, Bailey J, Eichele G, Lee CC (1997) RIGUI, a putative mammalian ortholog of the Drosophila period gene. Cell 90(6):1003–1011
23. Tosini G, Menaker M (1996) Circadian rhythms in cultured mammalian retina. Science 272(5260):419–421
24. Turek FW, Van Cauter E (1994) Chapter 43: rhythms in reproduction. In: Knobil E, Neill JD (eds) The physiology of reproduction, 2nd edn. Raven, New York, pp 487–540
25. Turek FW, Joshu C, Kohsaka A, Lin E, Ivanova G, McDearmon E, Laposky A, Losee-Olson S, Easton A, Jensen DR, Eckel RH, Takahashi JS, Bass J (2005) Obesity and metabolic syndrome in circadian clock mutant mice. Science 308:1043–1045
26. Turek FW (2008) Circadian clocks: tips from the tip of the iceberg. Nature 456(7224):881–883
27. Vitaterna MH, King DP, Chang A-M, Kornhauser JM, Lowrey PL, McDonald JD, Dove WF, Pinto LH, Turek FW, Takahashi JS (1994) Mutagenesis and mapping of a mouse gene, *Clock*, essential for circadian behavior. Science 264:719–725

Chapter 2
An Introduction to Chronobiology

Juan Antonio Madrid Pérez

Abstract One of the driving forces of evolution has been the constant search for independence from the uncertainties generated by a continuously changing environment. Faced with this selective pressure, organisms have developed mechanisms that permit uncertainties to be predicted, among which the biological clocks have been one of the most widely developed. By using biological clocks, organisms can anticipate periodical changes in their surroundings and adapt themselves to environmental changes that are cyclic in nature. In mammals, for example, the system charged with anticipating 24 h cycles is the circadian system. This is composed of a central pacemaker located in the suprachiasmatic nucleus of the hypothalamus (SCN), several peripheral oscillators; some inputs to the system, which act by synchronising the clocks to the environment, for example, the day–light cycle, feeding and physical exercise times; and outputs, which are all the rhythms that can be measured in an organism.

Besides permitting anticipation, the circadian rhythm maintains the internal coordination of all the biological rhythms of the different functions. This internal organisation is fundamental for maintaining health. However, the improper exposure to synchronizers and the loss of functionality associated with ageing favours the appearance of circadian dysfunctions, or chronodisruption. This impairment increases the incidence of certain pathologies, such as cognitive and affective disorders, accelerated ageing, metabolic syndrome, obesity, certain types of cancer, including colorectal, breast and prostate and sleep disorders, among others. Given the clinical importance of chronodisruption, techniques are needed to monitor circadian system status and procedures need to be developed for reinforcing its rhythmicity.

J.A. Madrid Pérez, Ph.D. (✉)
Chronobiology Laboratory, Department of Physiology, Faculty of Biology, University of Murcia, Murcia, Spain
e-mail: jamadrid@um.es

M. Garaulet and J.M. Ordovás (eds.), *Chronobiology and Obesity*,
DOI 10.1007/978-1-4614-5082-5_2,

Abbreviations

SCN	Suprachiasmatic nucleus
RTH	Retinohypothalamic tract
PER	Period homolog
BMAL1 or ARNTL or MOP3	Aryl hydrocarbon receptor nuclear translocator-like
CLOCK	Circadian locomotor output cycles kaput
CRY	Cryptochrome
TGFα	Transformant growth factor alpha
bHLH	Basic helix–loop–helix
PAS	Period-Arnt-Singleminded
CCGs	Clock controlled genes
CK1ε	Casein kinase 1 epsilon
CK1δ	Casein kinase 1 delta
Rev-Erbα or Nr1day	Nuclear receptor subfamily 1 group D, member 1
ROR	RAR-related orphan receptor
TAP	Algorithm based in the integration of temperature activity and position rhythms
ACM	Ambulatory circadian monitoring procedure
DLMO	Dim light melatonin onset
PCR	Polymerase chain reaction
RT-PCR	Real time-polymerase chain reaction
Q-PCR	Quantitative PCR
MESOR	Midline Estimating Statistic of Rhythm
IS	Interdaily stability
IV	Intradaily variability
L5	Least active 5 h
M10	Most active 10 h
AMP	Amplitude
RA	Relative amplitude
WT	Wrist skin temperature
NSAIDS	Non-steroidal anti-inflammatory drug

Introduction: The Circadian System

Due to its importance for species survival, natural selection has led to all organisms possessing a set of structures capable of generating and synchronising oscillations with periods of approximately 24 h in biological variables (known as a whole as circadian system). The existence of such a system capable of organising the rhythms of biological processes endows organisms with two major advantages. The first is that they can anticipate periodical and predictable changes in the environment, for example sunrise and sunset, the arrival of predators or the availability of food. The

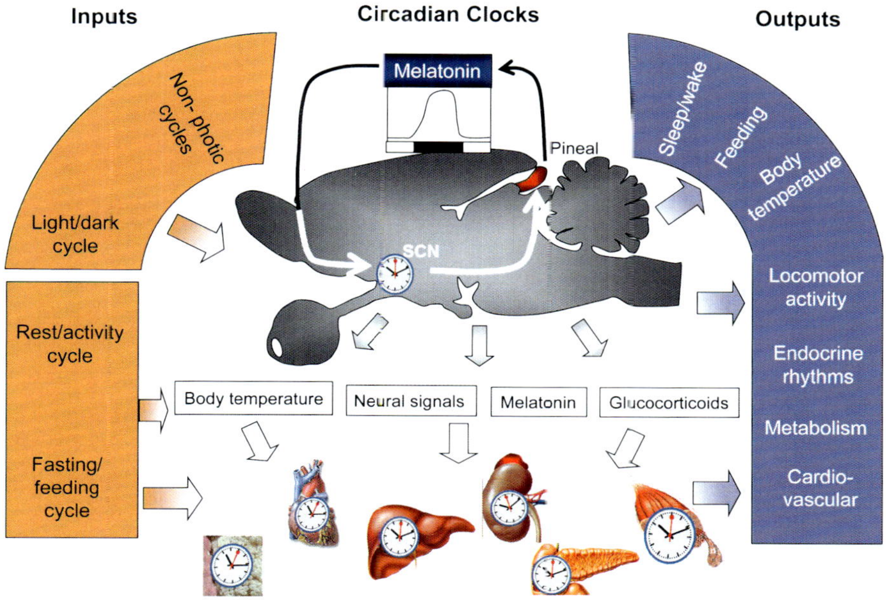

Fig. 2.1 General organisation of mammalian circadian system. *Inputs signals*: environmental cyclic cues can synchronise the activity of the central pacemaker and peripheral oscillators. *Circadian clocks*: the oscillatory machinery is composed of a *central pacemaker* and *peripheral oscillators*. The SCN is considered the major pacemaker of the circadian system, driving circadian rhythmicity in other brain areas and peripheral tissues by sending them neural, humoral and physical signals. Most peripheral tissues and organs contain circadian oscillators. Usually they are under the control of the SCN; however, under some circumstances (i.e. restricted feeding, jetlag, shift work, etc.), they can desynchronize from the SCN. *Outputs*: central pacemakers and peripheral oscillators are responsible for the daily rhythmicity observed in most physiological and behavioural functions. Some of these over-rhythms (physical exercise, core body temperature, sleep–wake cycle and feeding time), in turn, provide a *feedback*, which can modify the function of SCN and peripheral oscillators

second advantage is that it permits the temporal separation of antagonistic processes that take place in the same cell, for example, both lipolytic and lipogenic enzymes are present in the same hepatocytes, but they are active at different time [1]. If the activity of both enzyme types will increase at the same time, its effect on fat storage and metabolization would be completely ineffectual.

Although the molecular bases of the circadian clock of most species share common mechanisms, their anatomical and functional organisation differs between animals. The circadian system of mammals has three main components: circadian clocks, the input pathways, which participate in environmental synchronisation (or entraining) and the output pathways, which transmit the temporal circadian signals to the rest of the effector systems of the organism (Fig. 2.1).

Circadian Clocks

In mammals, the principal circadian clock or master pacemaker is located in the suprachiasmatic nuclei of the hypothalamus (SCN) [2, 3]. These are two small nuclei composed of several thousand neurons located each side of the third ventricle in dorsal position with respect to the optic chiasm. With their rhythmic activity, the SCN are responsible for most circadian rhythms of mammals, as revealed by lesion studies, which result in arrhythmicity in most of the variables recorded. In contrast, the transplantation of SCN from donor to arrhythmic SCN-lesioned animals restores the rhythmicity, confirming that the SCN is a circadian pacemaker in mammals [4].

Besides the principal hypothalamic pacemaker, the circadian system is composed of numerous secondary oscillators (cerebral cortex, liver, kidney, adipose tissue), which are capable of producing circadian oscillations and which in normal conditions are under the control of the SCN [5, 6]. However, these oscillators can also operate autonomously for several days in tissue cultures and can be synchronised by periodical signals other than light, such as temperature cycles or nutrient availability.

Input Pathways

The action of the SCN is sufficient for circadian rhythms to appear. However, if these rhythms are to be synchronised to environmental cycles, the circadian pacemaker has to be set periodically by the action of certain environmental factors (synchronizers or *zeitgebers*), which oscillate rhythmically and which, taken as a whole, act as the clock input pathway. Among these *zeitgebers* the most important is the light–dark cycle, although meal times, scheduled exercise, sleep and social contacts also act as synchronizers [1, 7–9].

The main input pathway of photic information to the SCN is the retinohypothalamic tract (RHT), formed by axons of a subpopulation of ganglion cells in the retina, which do not take part in conscious image formation. These cells contain a photopigment, melanopsin, which is not present in rods and cones [10]. The presence of these photosensitive cells, which project to the SCN, explains why the circadian rhythms of blind rats (without cones and rods) remain synchronised to the environmental light–dark cycle [11, 12].

Output Pathways

The output signals from the CSN transmit information to the regions of the brain that participates in the regulation of behaviour, sleep–wakefulness and body temperature, to the neuroendocrine centres and to peripheral organs. To this, the SCN use neural projections, humoral mediators, such as melatonin and cortisol, and physical signals like the rhythm of the central temperature [13–15].

The SCN projects axons to a several hypothalamic regions, the preoptic region, the anterior brain and the thalamus [14]. There are also direct connections with autonomous neurons that selectively transmit temporal signals to different organs (liver, pancreas) and endocrine glands (testes, adipose tissue, suprarenal gland) [16].

The fact that transplantation of the SCN was capable of restoring the rhythmicity of locomotor activity in lesioned animals, even when the implant was encapsulated in a semipermeable membrane that prevents neural growth [17], demonstrated that, besides nervous connections, the SCN, releases humoral mediators such as transformant growth factor (TGFα) [18], cardiotropin [19] and prokineticine-2 [20], which are involved in the transmission of temporal information.

One of the best characterised humoral mediators of the SCN is melatonin, whose synthesis is controlled by the SCN through a multisynaptic pathway that reaches the pineal gland after a relay in the upper cervical ganglia [21, 22]. This hormone is involved in the regulation of sleep, and both circadian and seasonal rhythms [23, 24]. Its synthesis is subject to a double regulation, on the one hand it respond to noradrenergic stimulation of the SCN and, on the other, to the direct inhibitory action of light [25].

The production of this hormone shows a marked circadian rhythm, with low plasma levels during the day and peaking during the night regardless of the nocturnal or diurnal characteristics of the organism [26]. Thus, melatonin is also known as "chemical darkness". The great stability of the melatonin cycle and the fact that it is produced during darkness means that it can be used by organisms as a daily clock, informing them o the arrival of night and as a calendar that tells them which season they are in [21, 27].

The Molecular Clock

Both in the SCN and in the peripheral oscillators each cell behaves as an autonomous circadian oscillator. At cell level, the circadian oscillators are the result of the existence of positive and negative feedback loops in which the products of the expression of given genes inhibit their own transcription, generating a rhythmicity of around 24 h (Fig. 2.2) [28]. The main components that have been identified in the clock of mammals are: the *Clock* and *Bmal1* genes as positive elements, and the *Per* genes (*Per1*, *Per2* and *Per3*) and the cryptochromes (*Cry1* and *Cry2*) as negative elements [29].

CLOCK and BMAL1 proteins are transcription factors that possess the functional domain bHLH ("*Basic Helix–Loop–Helix*") which confer them with a DNA binding capacity. These proteins heterodimerize in the cytoplasm through PAS domains (a name derived from the three proteins they share, *Period-Arnt-Singleminded*) and translocate to the nucleus, where they activate the transcription of given target genes (*Per*, *Cry*, *Rev-Erbα*) and clock controlled genes (CCG, including

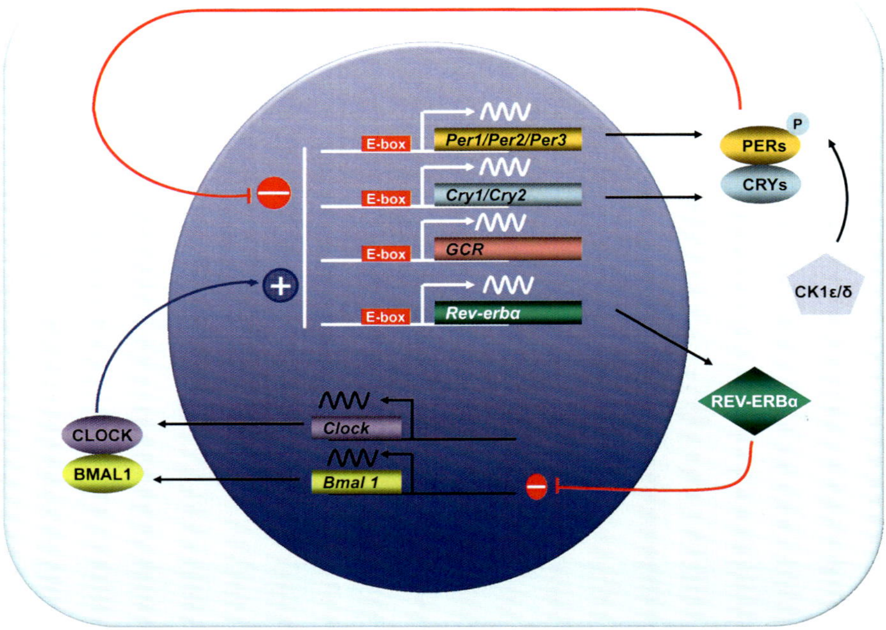

Fig. 2.2 Organisation of the molecular mammalian oscillator. The cellular oscillatory machinery is composed of a positive (CLOCK and BMAL1) and a negative (PER1–3 and CRY1, 2) limb. CLOCK–BMAL1 heterodimers, after binding to E-box elements, drive the transcription of several genes: *Cry1–2, Per1-3, Rev-Erbα* and multiple clock controlled genes (CCGs). After dimerization PERs and CRYs undergo nuclear translocation inhibiting CLOCK–BMAL1-mediated transcription. When the levels of PERs and CRYs fall, the negative repression is lifted and CLOCK–BMAL1 activates again gene expression. A secondary loop is established by the negative, REV-ERBα effect on *Bmal1* transcription. The molecular circadian clock drives the expression of several genes, CCGs, responsible of the generation of an internal temporal order in physiological, biochemical and behavioural rhythms

key regulators of the cell cycle and metabolism). The CCG make up about 10–30 % of the genome, depending on the tissue. The mechanism that permits the connection between the clock genes-proteins and the CCG is the binding to the *E-box* in the consensus sequence CACGTG.

The negative feedback loop comprises the heterodimers PER:CRY which translocate to the nucleus where they suppress their own transcription by inhibiting CLOCK and BMAL1 activities. Meanwhile, the protein REV-ERBα suppresses *Bmal1* transcription by binding to the elements conforming the response to Rev-erb*α*/ROR present in their promoter. Consequently, the RNA levels of *Bmal1* diminish, while those of *Per* and *Cry* increase. When the heterodimers CRY:PER suppress their own transcription at nuclear level (through acting on CLOCK-BMAL1), they also inhibit the transcription of *Rev-erbα*, permitting the transcription of *Bmal1* to be activated.

The approximately 24 h rhythmicity of the molecular clock mainly derives from post-translational modifications such as phosphorylation and ubiquitination, processes which affect the stability and translocation of the clock genes to the nucleus [30]. In this way, the casein kinase 1 epsilon (CK1ε) and casein kinase 1 delta (CK1δ) are the critical factors that modulate the functioning of the clock.

The importance of these post-translational modifications of the clock components is borne out by studies showing that mutations in CK1ε have an effect on circadian periodicity [31]. CK1ε phosphorylates the PER proteins, so that they are not immediately available to form dimers, leading to a circadian cycle with a longer period. When this gene is mutated so that the protein has reduced phosphorylation activity, the PER proteins are internalised into the nucleus more rapidly, shortening the cycle. In humans, the advanced sleep phase syndrome has been associated with a mutation in the gene *Per2*, which generates a mutated protein that cannot be phosphorylated efficiently by CK1ε, leading to the more rapid accumulation of PER2 and a shorter molecular cycle [32]. In the case of delayed sleep phase syndrome, a significant correlation has been found with a given polymorphism of the gene *Per3*, although the mechanism by which this change in the *Per3* sequence can determine a delay in the clock phase is unclear [33]. The morning or evening chronotype of an individual may also have a genetic basis, and it has been suggested that a polymorphism observed in the *Clock* gene may explain the difference between both chronotypes [34].

The discovery of new clock genes and of genes whose expression is controlled by the clock should permit the genetic bases of the syndromes that involve the circadian clock to be established.

Some Basic Issues in Chronobiology

Chronobiology (from the Greek Kronos = time; bios = life and logos = science) is the scientific discipline that studies rhythmic changes (biological rhythms) in living organisms at their different levels of organisation. A biological rhythm is the recurrence of a biological phenomenon at regular time intervals [35]. Among the parameters that characterise biological rhythms, the period or time that elapses for a complete oscillation to be repeated and its inverse, the frequency or number of cycles per unit of time, are widely used. In chronobiology, the most frequently used unit of frequency is the day. In this way, rhythms can be classified as circadian, with a frequency close to 1 day (between >20 and <28 h), ultradian, with a frequency greater than a cycle per day (<20 h) and infradian, rhythms with a frequency of less than one cycle per day (>28 h). Among the last of these, are included circalunar (≈28 days), circannual (365 days) and circaseptan (7 years) rhythms.

One of the principal properties of biological rhythms is that they persist in the laboratory under constant environmental conditions (free running), that is, they have an endogenous character [36]. However, for rhythm to maintain a given phase relation with the environment, the circadian system must possess synchronisation mechanisms permitting them to adjust to the environmental cycles. This property, known as synchronisation capacity, automatically corrects the delays or advances produced daily in clock functioning [35]. For an environmental factor to act as synchroniser or *zeitgeber*, its period must be very stable, so that it is unsurprising that the main synchroniser is the light–dark cycle. A third property of biological rhythms is the fact that the circadian clock does not alter significantly when it is measured at different environmental temperatures (temperature compensation) [37].

Monitoring the Functional Status of the Circadian System

The role placed by the circadian system in maintaining health underlines the importance of developing techniques for its objective evaluation, just as there are techniques that evaluate the respiratory and cardiovascular systems. The main challenge is to be able to measure a process that develops over long periods of time, which implies multiple measurements, preferably ones that do not interfere with the subject's daily routine [38]. The most convenient in this respect is to measure the outputs (circadian marker rhythms), as one does with the hands of a clock. Among such outputs, the rhythm of body temperature, motor activity, melatonin, cortisol and clock gene expression are the most commonly used [39]. In addition to these techniques, self-reported questionnaires based in the morning-evening preferences have been developed as complementary procedures to assess human chronotypes [40–43].

Central and Peripheral Thermometry

One of the most commonly used marker rhythms is the central temperature rhythm, whose profile has been widely described [44–46], and in which the highest values occur in the day and lowest at night.

In humans, the central temperature is usually measured by means of rectal probes that should be worn for several days, which is obviously uncomfortable. Recently as an alternative to measuring the central temperature, the rhythm of skin peripheral temperature has been proposed as a marker rhythm [47, 48]. This rhythm is induced by the alternation between vasodilatation and vasoconstriction generated by the parasympathetic–sympathetic balance. The predominance of sympathetic activity during the day is associated with lower temperatures, while its inhibition and the simultaneous activation of the parasympathetic system are associated with higher temperature. Moreover, increased skin temperature constitutes a signal that favours the beginning of nocturnal sleep through stimulation of hypothalamic areas [49].

The circadian pattern of peripheral skin temperature exhibits some characteristics phases (Fig. 2.3) [47]. It increases prior to sleep and remain high during the night. Upon awakening, the temperature falls abruptly and remains low during the day. About 20–21 h, when the peripheral temperature reaches its lowest value it is

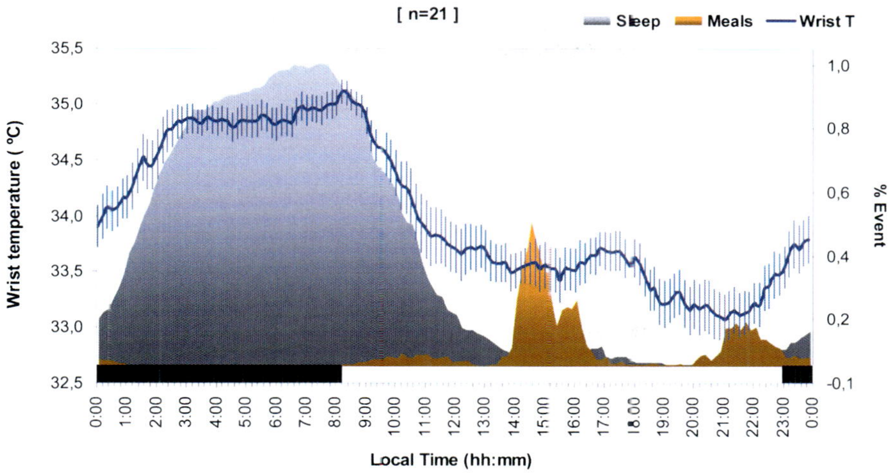

Fig. 2.3 Mean 24-h waveform of wrist skin temperature (WT) of young healthy students ($n=21$) recorded continuously for 7 days during a holiday week. WT (*blue line*), percentage of subjects asleep at any given time (*shaded grey area*), and time of feeding (*solid brown area*). *Black* and *white bar* indicates the sleep period of reference (from 23:00 to 08:00 h). The standard skin temperature rhythm exhibits a sharp increase in anticipation to sleep onset, it maintains high levels during nocturnal sleep and shows a secondary peak in the afternoon. More information in Sarabia et al. 2008, [47]

difficult to go to sleep in normal circumstances, this phase is known as wake maintenance zone.

The most used procedure to record skin temperature consists of a small autonomous data logger, placed on the internal surface of the wrist (over a radial artery) of the non-dominant hand and held in place by a bracelet or watch. The sensor can also be placed in any other peripheral region such as the arm, ankle or finger [47, 48].

Actimetry

A much more widely used method, especially in clinical practice, is actimetry. This technique consists in the use of a device that registers the displacement of the body zones where it is installed at frequent and regular intervals. Actigraphy is a non-invasive method useful to measuring the rest–activity cycle in humans. It is based on the principle that during periods when the individual is awaken, activity levels are high compared to when the individual is asleep. For its measurement, an activity sensor (actimeter) is placed on the wrist of the non-dominant hand for not less than 5 days, the minimum period to obtain reliable data that reflect the characteristics of the subject [50].

Actimetry is considered the method of choice for evaluating and diagnosing circadian disorders such as chronodisruption in shift-workers, delayed and advanced sleep phase syndrome, free running syndrome and irregular circadian rhythms [51]. However, as with any other measurement, actimetry is subject to masking and artefacts, for example, the difficulty in differentiating between the beginning of night rest and the removal of the sensor to shower just before going to bed, movements of one's bed partner, sleeping in a car or train, etc. [48, 50, 52].

Light Exposure

The circadian system is regulated by external signals, which are responsible for setting the clock each day. Given that the light–dark cycle is the most important synchroniser, it is of great interest to be able to quantify the light exposure of individuals. Low levels of illumination during the day lower the central temperature and state of awareness compared with the levels observed in high illumination [53–55]. However, during the night exposure to light, especially blue light, should be avoided in order to maintain melatonin secretion.

The light–darkness cycle to which subjects are exposed can be quantified by small data loggers, that contain a photosensitive cell, that periodically record the light intensity received by the individual [55]. Recently, sensors that differentiate between light wavelengths (blue, red and green) have become available, which means that the blue light, which has a greater capacity to synchronise the circadian pacemaker, can be accurately evaluated. The combination of these sensors with environmental temperature sensors provides complete information concerning the quality of the environmental synchronisers that act on the circadian system [55]. They also enable poor sleep hygiene habits to be identified; for example, sleeping in illuminated environmental conditions or in too high temperatures.

Integration of Variables

In order to increase the reliability of circadian monitoring, integrated variables obtained from processing individual variables have been recently proposed. For example, the TAP algorithm, proposed by Ortiz-Tudela et al. [48], is based on integrating, after normalisation, the following variables: skin temperature, motor activity and body position (Fig. 2.4). The first of these variables, skin temperature, is under endogenous control, while motor activity is modified voluntarily but it is also under endogenous control. Lastly, of the three variables used for the integration, body position is the most closely dependent on voluntary control. TAP is modular thus it can be amplified by incorporating new variables that complement the information even further. TAP variable permits us not only to determine how the individual's circadian system functions, but also to infer the sleep–wake rhythm with a precision higher than 90 % according to polysomnographic recording. This technique constitutes the base of ambulatory circadian monitoring procedure (ACM)

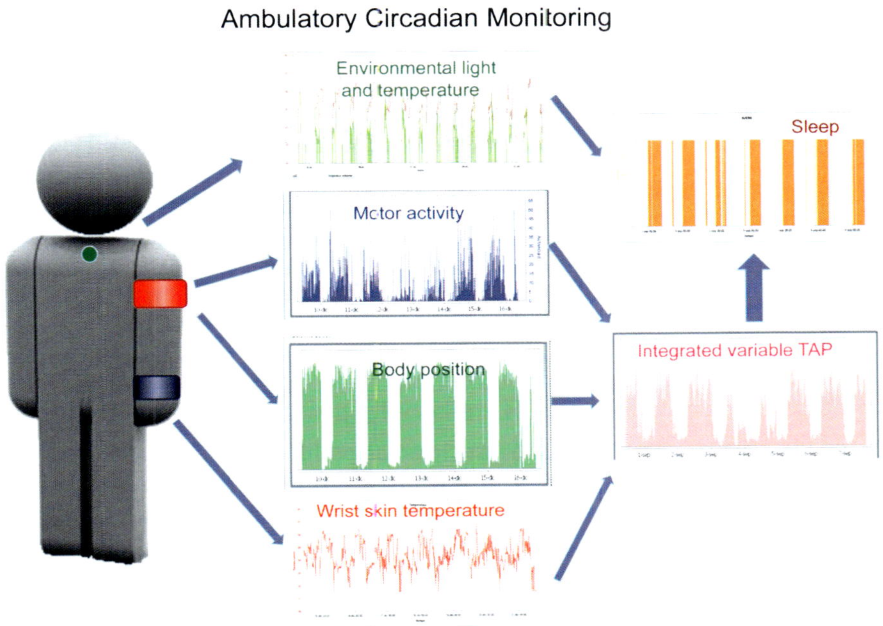

Fig. 2.4 Protocol for the assessment of functional status of the circadian system, using ambulatory circadian monitoring. Data of environmental light and temperature, motor activity, body position and wrist skin temperature are recorded by means of three sensors placed around the neck, arm and wrist. The integrated variable TAP is obtained from the normalisation of data from skin temperature, activity and position using appropriate algorithms. From TAP data, sleep–wake rhythm can be deduced with high accuracy. More information in Ortiz-Tudela et al. 2010, [48]

(Fig. 2.4), which recently has been applied to evaluating the circadian maturity in newborns [56], and pathologies like metabolic syndrome [57].

Melatonin

Melatonin is considered the best marker of the circadian system phase. However, its profile is strongly influenced by light exposure and, to a lesser extent, body position, physical activity, sleep, caffeine and drugs like NSAIDS and beta-blockers [58–64]. Plasma levels of melatonin show a circadian profile, with low levels during the day and high levels during the night, the highest being between 02:00 and 04:00 a.m. In humans, melatonin contributes to the body temperature rhythm since it is responsible for vasodilatation of the skin of the extremities through its activation of thermosensitive neurons present in brain areas involved in sleep regulation. The melatonin secretion schedule is closely related with the propensity to sleep and coincides with a fall in the central body temperature, arousal level and performance [65].

68. Slag MF, Ahmad M, Gannon MC, Nuttall FQ (1981) Meal stimulation of cortisol secretion: a protein induced effect. Metabolism 30:1104–1108
69. Van Coevorden A, Mockel J, Laurent E, Kerkhofs M, L'Hermite-Balériaux M, Decoster C, Nève P, Van Cauter E (1991) Neuroendocrine rhythms and sleep in aging men. Am J Physiol 260:E651–E661
70. Caufriez A, Moreno-Reyes R, Leproult R, Vertongen F, Van Cauter E, Copinschi G (2002) Immediate effects of an 8-h advance shift of the rest-activity cycle on 24-h profiles of cortisol. Am J Physiol Endocrinol Metab 282:E1147–E1153
71. Johnson CH (1999) Forty years of PRCs—what have we learned? Chronobiol Int 16:711–743
72. Bjarnason GA, Jordan RC, Wood PA, Li Q, Lincoln DW, Sothern RB, Hrushesky WJ, Ben-David Y (2001) Circadian expression of clock genes in human oral mucosa and skin: association with specific cell-cycle phases. Am J Pathol 158:1793–1801
73. Cajochen C, Jud C, Münch M, Kobialka S, Wirz-Justice A, Albrecht U (2006) Evening exposure to blue light stimulates the expression of the clock gene PER2 in humans. Eur J Neurosci 23:1082–1086
74. Minors DS, Waterhouse JM (1984) The use of constant routines in unmasking the endogenous component of human circadian rhythms. Chronobiol Int 1:205–216
75. Waterhouse J, Jones K, Edwards B, Harrison Y, Nevill A, Reilly T (2004) Lack of evidence for a marked endogenous component determining food intake in humans during forced desynchrony. Chronobiol Int 21:445–468
76. Halberg F (1970) Frequency spectra and cosinor for evaluating circadian rhythms in rodent data and in man during Gemini and Vostok flights. Life Sci Space Res 8:188–214
77. van Someren EJ, Hagebeuk EE, Lijzenga C, Scheltens P, de Rooij SE, Jonker C, Pot AM, Mirmiran M, Swaab DF (1996) Circadian rest-activity rhythm disturbances in Alzheimer's disease. Biol Psychiatry 40:259–270

Chapter 3
Adipose Tissue as a Peripheral Clock

Purificación Gómez-Abellán and Marta Garaulet

Abstract Adipose tissue is a complex and highly metabolic and endocrine organ, which is capable of expressing and secreting a variety of bioactive peptides, so-called adipokines. These adipokines are involved in coordinating a diversity of biological processes including energy intake and expenditure, insulin resistance, adipocyte differentiation, dyslipidemia, and body fat distribution. One of the most outstanding discoveries in the last year is the presence of an active circadian clock in adipose tissue depots. New data suggest that there is a temporal component in the regulation of all these adipose tissue functions. In fact, studies performed by microarrays have shown that a certain percentage of active genes expressed in adipose tissue in both humans and animal models follow a daily rhythmic pattern. Examples of these genes are clock genes (*PER2, CLOCK, CRY1,* and *BMAL1*), adipokine genes (adiponectin and leptin), and glucocorticoid-related genes among others. Thus, an adequate temporal order in the daily pattern of these genes implicated in adipose tissue metabolism could have important consequences not only in body fat distribution but also in the metabolic alterations associated to obesity. Further investigations of circadian rhythms in adipose tissues will provide insight into the physiology of energy homeostasis and the etiology of metabolic diseases such as obesity.

Abbreviations

LEP	Leptin
LEPR	Leptin receptor
TNFα	Tumour necrosis factor alpha

P. Gómez-Abellán, Ph.D. (✉) • M. Garaulet, Ph.D.
Department of Physiology, University of Murcia, Murcia, Spain
e-mail: puriki4@hotmail.com; garaulet@um.es

M. Garaulet and J.M. Ordovás (eds.), *Chronobiology and Obesity*,
DOI 10.1007/978-1-4614-5082-5_3,

ApM1	Adipose most abundant gene transcript 1 or adiponectin
Acrp30	Adipocyte complement-related protein of 30 kDa or adiponectin
GBP28	Gelatin binding protein of 28 kDa or adiponectin
AdipoQ	Adiponectin
ADIPOR1	Adiponectin receptor 1
ADIPOR2	Adiponectin receptor 2
TZDs	Thiazolidinediones
IL-6	Interleukin 6
ASP	Acylation stimulating protein
PAI-1	Plasminogen activator inhibitor-1
TGF-beta	Transforming growth factor-beta
GCs	Glucocorticoids
HPA	Hypothalamic–pituitary–adrenal
GR	Glucocorticoid receptor
11DHC	11-Dehydrocorticosterone
11βHSD1	11β-Hydroxysteroid dehydrogenase 1
11βHSD2	11β-Hydroxysteroid dehydrogenase 2
STAR	Teroidogenic acute regulatory protein
5αR	5-α reductase
PPARγ	Peroxisome proliferator-activated receptor gamma
PCR	Polymerase reaction chain
AT	Adipose tissue
WAT	White adipose tissue
BAT	Brown adipose tissue
SWAT or SAT	Subcutaneous adipose tissue
VWAT or VAT	Visceral adipose tissue
IAAT	Intra-abdominal adipose tissue
FFAs	Free fatty acids
TGs	Triglycerides
MetS	Metabolic syndrome
SCN	Suprachiasmatic nucleus
LPL	Lipoprotein lipase
Pdp1	PAR domain protein 1
RORα	RAR-related orphan receptor alpha
PGC1α	Peroxisome proliferative activated receptor gamma, coactivator 1 alpha
PER2	Period homolog 2 (Drosophila)
BMAL1 or ARNTL or MOP3	Aryl hydrocarbon receptor nuclear translocator-like
CLOCK	Circadian locomotor output cycles kaput
CRY	Cryptochrome
mRNA	Messenger ribonucleic acid
CCG	Clock control genes
ASCs	Adipose-derived stem cells
ACTH	Adrenocorticotropic hormone

Adipose Tissue as Endocrine Organ

Classically, adipose tissue has been regarded as a passive reservoir for energy storage, but this traditional point of view is no longer valid. In 1987, adipose tissue was identified as a major site for metabolism of sex steroids [1]. Nevertheless, the critical change in our perspectives on adipose tissue came in 1994 with the discovery of the cytokine-like factor, *leptin* (from the Greek *leptos*, meaning thin). For the first time in science, it was described that adipose tissue was able to secrete hormones or "adipocytokines" capable of communicating information from periphery to the central nervous system [2]. This outstanding outcome opened a "great window" in the study of obesity. Adipose tissue was not any more a passive reservoir of energy; it was as an endocrine organ and with this discovers a new "époque" started in the obesity research.

Nowadays we know that adipocytes secrete leptin in direct proportion to adipose tissue mass as well as nutritional status, in this way leptin signals the status of energy stores and its secretion can reduce appetite and increase energy expenditure. The capability of leptin to regulate food intake, body weight, and adiposity has been recognized entirely to its actions in the hypothalamus [3]. Nevertheless, it has been reported that leptin is essential in the adipose tissue itself, modulating the adipocytes' metabolic function, up-regulating fat oxidation, and decreasing lipogenesis [4]. These physiological functions are carried out by binding to its receptor (*LEPR*) expressed in adipose tissue [5]. Therefore, changes in leptin or its receptor in adipose tissue can be relevant in the development of obesity and other metabolic disorders [6].

Since the discovery of leptin, numerous research findings show that adipose tissue is a highly active endocrine organ, which is involved in many physiological processes. These metabolic processes are influenced by products of the adipose tissue, so-called adipocytokines or adipokines (Fig. 3.1).

One of the first molecules studied in relation to adipose tissue was *TNFα*. This particular protein was firstly analyzed as an inflammatory molecule in cancer studies.

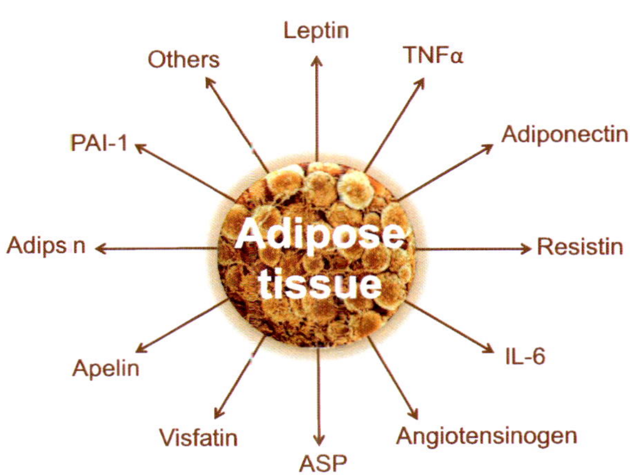

Fig. 3.1 Adipose tissue as an endocrine organ. AT can secrete many adipokines and certain factors involved in the metabolism such leptin, TNFα, adiponectin, visfatin among others

Within adipose tissue, TNFα is expressed by adipocytes and stromovascular cells [7] and nowadays we know that this cytokine is able to repress genes involved in uptake and storage of nonesterified fatty acids and glucose; it also suppresses genes for transcription factors involved in adipogenesis and lipogenesis; it changes the expression of several adipocyte-secreted factors including adiponectin and is capable of impairing insulin signaling [8, 9]. In this context, TNFα production is increased in obesity and it has been implicated in the development of insulin resistance in the adipocyte of the obese by altering insulin signaling through an autocrine or paracrine action.

In relation to insulin resistance, another important cytokine that has attracted much attention in the last years is *Adiponectin* [10]. This cytokine has been described as one of the most protective molecules against the different alterations associated with obesity, such as insulin resistance and inflammation Adiponectin was firstly characterized in 1995, and because it is one of the most expressed genes in adipose tissue it was formerly called apM1 (adipose most abundant gene transcript 1). The identification of adiponectin in the text is not easy and may be confusing, because it has been also named as Acrp30 (adipocyte complement-related protein of 30 kDa), adipoQ, and GBP28 (gelatin binding protein of 28 kDa) [10–13].

Several metabolic effects from adiponectin have been described. On the one hand, its *anti-atherogenic effect* is well known since it is capable of inhibiting expression of adhesion molecules and vascular smooth muscle cells proliferation, and suppresses transformation of macrophages to foam cells. Moreover, it is an *anti-diabetic cytokine* because increases insulin sensitivity and decreases hepatic glucose output and it increases fatty acid oxidation. Besides, adiponectin is decreased in abdominal obesity [14].

The beneficial effects of this hormone are predominantly mediated by binding to two cell membrane receptors, adiponectin receptor 1 (*ADIPOR1*) and adiponectin receptor 2 (*ADIPOR2*) [15]. Both receptors are also present in adipose tissue, suggesting that adiponectin may have biological effects in adipose tissue in an autocrine/paracrine manner [16].

One of the most controversial adipokines described has been *Resistin* (resistance to insulin). This adipokine was identified in 2001 as a novel mRNA induced during adipocyte differentiation but down-regulated by thiazolidinediones (TZDs) in vitro [17]. Initial studies suggested that resistin had significant effects on insulin action, potentially linking obesity with insulin resistance and type II diabetes [18]. Biological activities of resistin include: (a) impairment of glucose tolerance in mice in vivo, (b) antagonism of glucose uptake by cultured 3T3-L1 adipocytes, and inhibition of 3T3-L1 differentiation into adipocytes [19]. Of note, in human studies, the influence of resistin on the development of insulin resistance is controversial. In obese subjects, resistin in adipose tissue is significantly higher compared to normal weight subjects [20, 21]. However, several human studies have failed to demonstrate any impact of obesity and insulin resistance on the concentration of resistin [22, 23].

Interleukin-6 (*IL-6*) is another cytokine associated with obesity and insulin resistance [24]. IL-6 comes from adipose tissue and this amount increases proportionally with increasing body mass [25]. This cytokine decreases the expression of insulin receptors in peripheral tissues, acts as an inhibitor of adipogenesis and inhibits adiponectin secretion. There is growing evidence that IL-6 and several other pro-inflammatory cytokines are "sleep factors". In addition, it has been shown that these cytokines also influence energy intake by enhancing insulin and leptin sensitivity. The study of this cytokine offers the possibility to connect the showed interactions among energy intake, sleep behaviour, and circadian system with metabolic alterations.

Human studies have indicated that complement proteins, secreted by adipocytes, the so-called *Complement-Related Proteins*, positively correlated with adiposity, insulin resistance, dyslipidemia, and cardiovascular disease [25]. *Adipsin* (complement factor D) is one of several adipose tissue-derived complement components that are required for the enzymatic production of acylation stimulating protein (*ASP*), a complement protein that affects both lipid and glucose metabolism [26, 27].

Every day we have new cytokines in research. In fact, *Visfatin* is a recently discovered adipokine produced and secreted primarily by visceral adipose tissue, which binds to and activates the insulin receptor, exerting insulin-mimetic effects both in vitro and in vivo [28]. There are many other adipokines, such as *Apelin*, which is an adipocytokine whose plasma concentration is increased in obesity, insulin resistance, and hyperinsulinemia [29]. But also other factors that are secreted by adipose tissue include angiotensinogen, plasminogen activator inhibitor-1 (PAI-1), tissue factor and transforming growth factor-beta (TGF-beta), adipophilin, monobutyrin, agouti protein, and factors related to pro-inflammatory and immune processes [30]. Although adipose tissue-derived hormones are identified every day, even those factors that were already characterized, such as leptin, require precise definitions of their physiological effects, as many as 40% are novel genes [31]. The continued identification characterization of these novel genes is likely the endocrine function of adipose between energy homeostasis and systems.

Glucocorticoids: The Circadian Hormone

One of the most outstanding circadian hormones in our body is cortisol. The analyses of its rhythms in plasma are one of the outcomes more frequently described in the classical chronobiology. It has been long established as a catabolic in nature, liberating energy substrates during times of stress to supply the increased metabolic demand of the body. The link between glucocorticoids (GCs) and excess adiposity is clearly demonstrated and well established clinically. However, the effects of GCs on adipose tissue metabolism are still contradictory. In fact, several studies have shown that the patients with elevated GCs, for example, individuals with Cushing syndrome [32] or those on exogenous corticosteroid treatment [33] present increased weight gain and visceral adiposity and are at increased risk for developing type 2 diabetes mellitus [34, 35]. Although elevated GC levels seem to contribute to visceral fat accumulation, most obese individuals do not exhibit elevated peak plasma GC levels [36–38]. Indeed, circulating GCs levels in obese patients might even be lower than in patients with normal weight, and intra-adipose GCs metabolism has been hypothesized as the reason for this low GCs plasma concentration [39]. On the other hand, subjects with visceral obesity show perturbations of the cortisol diurnal rhythm, characterized by a significant decrease in cortisol variability suggesting that a pathological HPA axis response is associated with abdominal fat distribution [40].

In this context, increased adipose tissue glucocorticoid exposure relies not only on glucocorticoid receptor (*GR*) availability but also on the local enzymatic interconversion of active (cortisol in humans and corticosterone in rodents) and inactive (cortisone in humans and 11-dehydrocorticosterone (11DHC) in rodents) hormones. This interconversion is controlled by two isoenzymes of 11β-hydroxysteroid dehydrogenase (*11βHSD1* and *11βHSD2*). But other enzymes can increase cortisol availability, such as steroidogenic acute regulatory protein (*STAR*), a key factor in steroidogenesis mediating the transfer of cholesterol from the outer to the inner mitochondrial membrane.

By contrast, cortisol can be inactivated by other enzymes, for example steroid 5-α reductase (*5αR*), an A-ring reductase that enhances cortisol clearance in peripheral tissues, mainly the liver [41–43]. In addition, the action of these genes can be regulated upstream by other factors which are crucial for the metabolism of adipose tissue such as *PPARγ*. This gene is an adipocyte-specific nuclear hormone receptor. Agonists of PPAR gamma, such as TZDs, promote adipocyte differentiation and have insulin-sensitizing effects [44]. Moreover, this is well established as *11βHSD1* activity is regulated by *PPARγ* [45], and it is known that some of the beneficial effects of *PPARγ* antidiabetic agents may result, at least in part, from the down-regulation of *11βHSD1* expression in adipose tissue [46]. The study of cortisol, and glucocorticoid receptors in adipose tissue, and their diurnal fluctuations is an important issue in obesity, from a chronobiological point of view, and perhaps could help us to clarify the important associations between stress, chronodisruption, and abdominal obesity.

Brown and White Adipose Tissue: Adipocytes or Stromal-Vascular Cells?

Although when we refer to adipose tissue, we consider it as a "whole", it is important to know that in adult mammals, the major volume of adipose tissue is a loose association of lipid-filled cells called adipocytes, which are held in a structure of collagen fibres. However, apart from adipocytes, adipose tissue contains stromal-vascular cells including fibroblastic connective tissue cells, leukocytes, macrophages, and adipocyte precursor cells, known as pre-adipocytes (not yet filled with lipid). In addition a combination of small blood vessels, connective tissue matrix, nerve tissue, stromovascular cells, and immune cells are also present [25], functioning as an integrated unit. One of the most frequent problems in the study of adipose tissue is the difficulty in assessing from the genes studied which are expressed by adipocytes or by the stromal vascular cells, and technique and histochemical and PCR techniques are helping us to clarify this aspect.

In mammals, mature adipocytes exist as two cytotypes, white and brown adipocytes, which are histologically distinct [47]. Whereas white adipocytes is characterized by

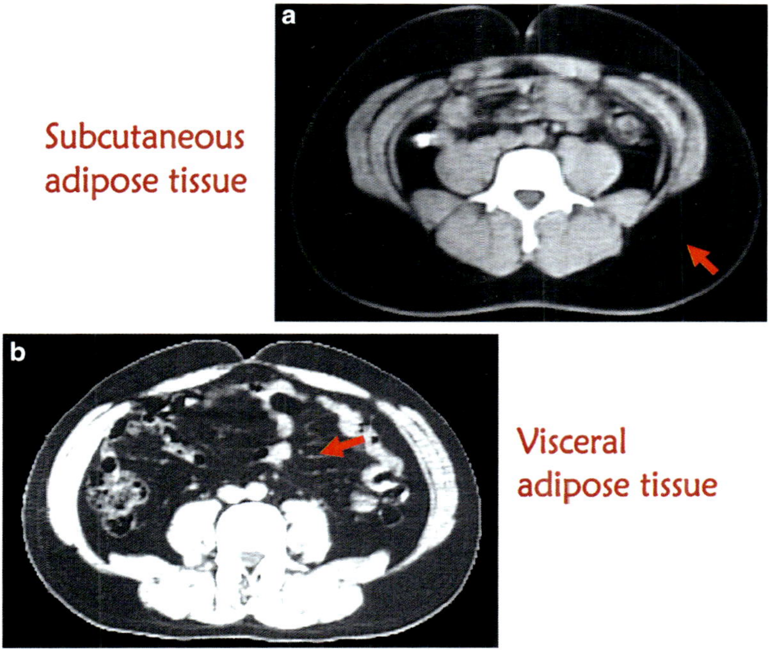

Fig. 3.2 A representation of two computed tomography slides. Subcutaneous fat-type with prominence of subcutaneous AT (**a**). Visceral fat-type: excess fat in the visceral compartment (**b**)

possessing lipids organized within one large and "unilocular" droplet which occupies the majority of intracellular space, compressing the cytoplasm and nucleus into a thin visible rim [48], brown adipocytes are organized into multiple smaller "multilocular" droplets and they are additionally characterized by their high content of large mitochondria packed with cristae within the cytoplasm. Moreover, brown adipocytes are polygonal, have centrally placed nuclei, and are relatively smaller than white adipocytes [49].

In this regard, mammals have two main types of adipose tissue: white adipose tissue (WAT) which contains white adipocytes, and brown adipose tissue (BAT) with brown adipocytes, each of which has different properties. While WAT is specialized in storing energy and is an important endocrine organ involved mainly in the control of weight regulation, the BAT is the main tissue regulating thermogenesis in response to food intake and cold [50]. Particularly, the WAT can be classified, depending on its distribution throughout the body in two major types: subcutaneous adipose tissue (SWAT or SAT) and visceral adipose tissue (VWAT or VAT) also called intra-abdominal adipose tissue (IAAT) (Fig. 3.2). SAT accumulates under the skin (known as peripheral fat mass) whereas VAT is located in the body cavity beneath the abdominal muscles and surrounding the intra-abdominal organs (well known as central fat mass).

Visceral and Subcutaneous, Two Different Adipose Tissues

Several studies have demonstrated that both depots, visceral and subcutaneous, are structural, physiological and metabolically different [51, 52]. Indeed, the type of fat cells or adipocytes, their endocrine function, lipolytic activity and the response to hormones differ between both adipose tissues. From the structural point of view, visceral adipose tissue contains greater number of large adipocytes and higher vascularity and innervation, in contrast to subcutaneous fat that possesses small adipocytes. Moreover, there are regional variations in density, affinity, and signal transduction of several adipose tissue receptors. VAT shows elevated concentrations of glucocorticoid [53] and androgen receptors [54], whereas oestrogen has greater binding capacity in SAT [55]. Respect to adrenergic receptors, visceral fat presents an increased β_3-adrenoreceptor and α_2-adrenergic receptor sensitivity to catecholamine stimulation compared with subcutaneous depot [56]. On the other hand, attending to physiological and metabolic differences, adipocytes from visceral adipose tissue are insulin-resistant, whereas adipocytes from SAT become more insulin-sensitive [57]. In general, visceral fat cells are metabolically more active than subcutaneous adipocytes, being hyperlipolytics and characterized to have higher rate of insulin-stimulated glucose uptake in contrast to SAT adipocytes which are more avid in absorption of circulating free fatty acids (FFAs) and triglycerides (TGs), preventing their deposition in non-adipose tissue [58, 59].

All these functional differences may be related to their anatomical location [60] and various physiological, psychosocial and clinical factors influence the amount and distribution of the adipose tissue throughout the human body, including sex, age, ethnicity, diet physical activity, hormone levels, among others [52]. Indeed, visceral fat accumulation has been associated to cause impaired glucose metabolism, elevated blood pressure, and dislipidemia, and therefore it is considered to be a key player in the metabolic syndrome (MetS) [61]. Moreover, this body composition phenotype is linked to other pathological conditions including several malignancies including prostate and colorectal cancers [62, 63]. On the other hand, peripheral fat mass is negatively correlated with atherogenic risk factors and improvement cardiovascular risk profile [60].

The "When" in the Adipose Tissue: A Peripheral Clock

Introduction

In recent years, numerous studies have provided the molecular mechanisms governing the regulation of circadian rhythms by neurons in the suprachiasmatic nucleus (SCN) of the anterior hypothalamus in the brain, the master circadian pacemaker [64]. In its simplest form, the molecular clock work consists of autoregulatory transcriptional and translational feedback loops that have both positive and negative elements, the *clock genes* [65]. These genes, orchestrated by the SCN in due time, are necessary for creating and sustaining rhythms of 24 h. In addition, posttranslational mechanisms such as protein phosphorylation, affect stabilization, degradation, and subcellular localization of clock proteins, thus contributing to the molecular clockwork [66, 67]. However, one of the most interesting outcomes in the chronobiological field has been the discovery of different peripheral clocks in many tissues and organs [68]. Peripheral clocks appear to be regulating rhythmicity of at least 10% of the expressed genes within each tissue, and their finding has revolutionized our understanding of how circadian and metabolic networks overlap to regulate physiologic processes in the whole organism [69].

In this context, the current scientific literature is replete with investigations providing a revolution in the study of adipose tissue biology. Many genes in adipose tissue show circadian rhythmicity [70]. Microarrays studies have documented that approximately 25% in humans and 50% in animal models of active genes expressed in adipose tissue follow a daily rhythmic pattern [71, 72] and depending of the tissue, between 10 and 30% of the total genes is under the control of the circadian molecular clock [73]. Thus, in a study, performed in human blood plasma and saliva, it has been demonstrated that approximately 15% of all identified metabolites are under circadian control and this strong effect of the endogenous circadian clock on multiple human metabolic pathways was independent of sleep or feeding [74].

The Importance of "Time" in Metabolism of Adipose Tissue

Food Intake and Lipogenesis and Lipogenetic Processes

All discoveries discussed above are clear evidences that in the metabolism of the adipose tissue it is important not only to understand the "what" and the "how" but also the "when" of the metabolic processes. Moreover, since the antagonism of many of the metabolic processes that occur in adipose tissue, it is expected that not all occur simultaneously. For example, mammals show alternating cycles of *lipogenesis and lipolysis*. Specifically in humans during the night (low activity period), there is a predominance of lipolytic activity, responsible of the body fat utilization, which reduces the frequency of hunger signals and, in consequence, reduces the need for food. In contrast, during the day, *lipogenesis* predominates, in order to fulfill energy needs during the activity period [75]. Indeed, several studies performed in metabolome and proteosome, point out to the presence of circadian rhythmicity in approximately 75% of the total lipids having their acrophases early in the morning and at 12 h, around the time of food intake [74].

Taking in consideration all these metabolic processes, we can suggest that time is important in adipose tissue and this tissue must be synchronized with different organs and tissues implicated in the food intake processes in order to be able to accumulate fat or mobilize it in the proper time. For that reason we can speculate that in adipose tissue, the genes implicated in these processes display also circadian rhythmicity in their expression and temporally are very precisely organized. In this sense, Gimble et al. in a study performed in humans have described that if a meal occurs out of phase with lipoprotein lipase (LPL) expression levels, an adipocyte-derived circulating enzyme (responsible for clearing circulating triglycerides), the individual may be prone to store circulating FFA in ectopic tissues, producing lipotoxicity and as a consequence hepatic, muscular or pancreatic comorbidities and MetS [76]. For this reason the circadian deregulation correlates with increased risk of obesity and its comorbidities as cardiovascular disease, diabetes and insulin resistance among other.

Circadian Rhythmicity in Adipokines

Other evidence that suggest a close relationship between circadian rhythms and adipose biology could be the fact that 24 h rhythms have been reported in the plasmatic concentration of leptin and adiponectin in humans [77, 78], being both adipokines. For example, leptin plasma levels during day are variables [77]. Normally, its peak is coinciding with the inactivity phase. Thus, in diurnal animals, such as humans, plasmatic leptin is high during night, when appetite decreases, and low during the day, when hunger increases [79]. However, in nocturnal animals, such rodents, its peak is during the early to mid-light phase [80]. With respect to adiponectin, Gavrila et al. showed that the 24-h variations of serum adiponectin was nearly identical and followed those of cortisol, but were out-of-phase with leptin diurnal rhythms in healthy men [78].

Circadian clocks have been shown to be present in adipose tissue of experimental animals [81] revealing rhythmic expression of clock and adipokines genes, such as resistin, visfatin, and adiponectin [82]. Moreover, diurnal variations in the sensitivity of adipose tissue to adrenaline-induced lipolysis persist ex vivo, suggesting that the intrinsic nature of the adipocyte exhibits a diurnal variation [83]. There are multiple works that show how circadian clock regulates metabolism in the adipose tissue [84].

Adipocytes Differentiation and Adipogenesis

It is well known the implication of clock genes in the adipocytes differentiation as well as in the control of adipogenesis and lipid metabolism [85, 86]. In this sense it has been crucial the availability of genetic models of circadian disruption because they have provided us new opportunities to dissect the interrelationship of circadian and metabolic systems. In fact, experimental studies performed in clock genes knockout mice showed how embryonic fibroblast failed to differentiate into adipocyte, and loss of clock genes led to a significant decrease in adipogenesis and gene expression of some key adipogenic/lipogenic factors [86]. On the contrary, overexpression of clock genes in adipocytes resulted in an increased lipid synthesis activity [86]. Other studies performed in "fat body" of Drosophila showed that *PAR domain protein 1* (*Pdp1ε*), equivalent to mammalian *RORα*, modulated a circadian output gene linked to starvation and feeding [87, 88].

Other Key Nutrient Sensors

Additional key nutrient sensors that have been implicated in the cross-talk between circadian rhythms and metabolism are PPARγ and the coactivator PGC1α (PPARγ coactivator). PPARγ is rhythmically expressed and directly regulates *Bmal1* transcription, and mice lacking PPARγ exhibit reduced rhythmicity of clock gene expression.

All these data indicate the importance of circadian rhythms and time in adipose tissue metabolism of experimental animals.

The Adipose Tissue: A Peripheral Clock

One of the most interesting discovers in the last times, related to obesity is the existence of a peripheral clock in human adipose tissue. Thus, it have been recently reported that clock genes are expressed in both human subcutaneous and visceral fat at a certain time of day [89] and this expression was sex dependent [90, 91]. But in addition to the basal expression, it has been showed that both negatives (*PER2* and *CRY1*) and positives (*CLOCK* and *BMAL1*) clock genes, showed circadian rhythmicity in its expression and oscillated independently of the suprachiasmatic nucleus in both adipose tissue explants ex vivo for at least two circadian cycles after surgery in morbidly obese women [92, 93]. These findings show the presence of active circadian clock mechanisms in human adipose tissue, being defined as a peripheral circadian oscillator.

Otway et al. conducted the first study to evaluate human adipose tissue circadian gene oscillations in vivo using serial biopsies, documenting that the oscillatory mRNA profile for core circadian genes was independent of body mass index [94]. These data are in contrast to experimental animals as rodents, in whom obesity attenuates circadian genes amplitude [82].

The Clock Control genes

Clock genes in fat tissue are also capable of modulating other genes, the so-called *Clock Control Genes* (*CCG*), which are not directly involved in the clock machinery but are able to induct the expression of many target genes. In this respect our own research group have published that different genes implicated in adipose tissue metabolism display circadian expression [92]. This is the case for *PPARγ*, a nuclear transcriptor factor which stimulates adipocyte differentiation, or circulating levels of certain hormones and cytokines highly related to adipose tissue, including adiponectin, leptin, tumour necrosis factor-α, interleukin-6 and plasminogen activator inhibitor-1 (PAI-1). Thus, several studies have demonstrated that some of these factors regulated by *PPARγ* have displayed a strong circadian pattern. Among these, adipokine genes as adiponectin (*ADIPOQ*) which displays a protective role against MetS disturbances, and leptin (*LEP*) closely related to the intake control [97, 98]. Although also their receptors, *ADIPOR1* and *ADIPOR2*, and *LEPR* [97, 98]. Other genes which oscillate with a circadian rhythm in culture of adipose tissue, are the cortisol metabolism-related genes (*GR, 11βHSD1, 11βHSD2, STAR* and *5αR*), highly implicated in food intake and central accumulation of fat [99].

Techniques in the Assessment of a Peripheral Clock in adipose tissue

The study of circadian rhythms in human adipose tissue has several technical limitations. The quantity of fat needed to assess different points increases, together with the ethical reasons make difficult to assess circadian rhythmicity in adipose tissue in normal weight patients, so most of the studies are made in morbid obese patients. Still, there are varied experimental approaches that have been employed to study adipose tissue from the chronobiological point of view (Table 3.1).

(a) *In vivo*

- *Single time point analyses.*
  The easiest analysis is to take a biopsy of adipose tissue and analyze adipose tissue clock gene expression. This can be performed from subcutaneous fat by kneel aspiration, or during surgery where you can also access to visceral fat.
  This was the method initially used by our group, to assess for the first time the expression of clock genes in human adipose tissue and their correlations with metabolic syndrome characteristic. Although these experiments provide useful preliminary data linking regional adipose clock gene expression with metabolic status, they are nonetheless subject to the disadvantage that a single time point analysis does not permit interpretation of temporal changes in clock genes expression.
- *Serial sampling of biopsies*
  Two studies have now described temporal changes in gene expression using serial subcutaneous adipose biopsies. In the first of these studies, three biopsies were collected in the morning, afternoon and evening from individuals; using array analysis it was estimated that approximately 25% of the human adipose transcriptome undergoes diurnal regulation [71].

  In a subsequent study, diurnal gene expression in individuals who were lean, mildly obese or obese with type 2 diabetes has been analyzed [94]. Biopsies were then collected every 6-h across a 24-h period from the upper buttock region, containing metabolically active adipose tissue [100]. Robust rhythms in all clock genes measured were observed and also in genes that

Table 3.1 Techniques in human adipose tissue circadian rhythmicity

Method	Advantage	Disadvantage
(a) *In vivo techniques*		
Single time point analyses	Study of both subcutaneous and visceral adipose tissue in vivo	Difficult interpretation of temporal changes in clock genes expression
	Provide useful preliminary data linking regional adipose clock with metabolic status	
Serial sampling of biopsies	Analyses of in vivo rhythms	Difficult to check if the clock is independent of the SCN
	Useful as a general marker for human metabolic rhythms	Difficult to determine the circadian pattern because sampling is in general limited to two or three samplings
		Cannot compare subcutaneous and visceral fat
(b) *In vitro techniques*		
Adiopocyte cells	Study of both subcutaneous and visceral adipose tissue	Does not reflect in vivo state of cells
	Identification of endogenous adipose rhythms	
Adipose tissue explants	Study of both subcutaneous and visceral adipose tissue	Limited tissue/sampling resolution
	Identification of endogenous adipose rhythms	Difficult to relate results to in vivo physiology
(c) *Other techniques*		
Bioluminescence imaging of cell culture	Study of both subcutaneous and visceral adipose tissue	Need living cells
	Identification of endogenous adipose rhythms without sampling at different times of day	Can be done in a whole animal but not in human
	Report high temporal resolution from a small amount of tissue	

Table adapted from Johnston JD [115]

have been linked to both circadian and metabolic function. Surprisingly, and in contrast to data from a similar mouse experiment [82], no significant differences in gene expression between our three experimental groups were found.

These studies are rather important particularly to assess potential differences between obese and normal-weight patients. However, these studies cannot answer the question about the existence of a peripheral clock in adipose tissue taking into account that circadian variability could be related to the influence of the SCN, not to the presence of a peripheral clock in the adipose tissue.

Moreover, it is difficult to determine the circadian pattern because sampling is in general limited to two or three samplings.

A third limitation is that we cannot compare subcutaneous and visceral fat.

(b) *In vitro techniques*

In vitro techniques are useful to assess the presence of a peripheral clock, because if rhythmicity persists in vitro, it may indicate the presence of a peripheral clock in the tissue, which is isolated to the SCN influence. These studies can be performed in:

- *Adipocyte cells*
- Authors have studied rhythms in human adipocyte cells differentiated from adipose-derived stem cells (ASCs) [96]. The disadvantage of this procedure is that it does not reflect in vivo state of cells.
- *Adipose tissue explants*
- These experiments also allow the study of both adipose depots and in addition the identification of endogenous adipose rhythms. In our group demonstrated the existence of the peripheral clock in human adipocytes by demonstrating the following statements:

1. Adipose tissue expresses clock genes
2. Clock genes fluctuate during 24 h, even outside the body
3. Clock gene expression is related to the protein expression
4. Clock genes may influence the circadian variability of clock control genes (CCG) and outstanding genes in adipose tissue metabolism.

A single biopsy from both subcutaneous and visceral regions is split into explants which are cultured and then collected for analysis, typically at 6-hourly intervals over a 24 h period. However, the technique does also provide some disadvantages. Firstly, temporal resolution of the analysis is limited by the amount of tissue that can be surgically removed by the current laparoscopic techniques. Our own experience shows that at least 12 g of fat are needed of each depot to define the rhythms of 24 h of any gene with six sampling points. This is heightened when the study is conducted in lean subjects, which normally do not have much fat tissue. Secondly, by moving the tissue into an in vitro environment, it is difficult to relate results to in vivo physiology.

Another important consideration when studying adipose tissue is its heterogeneous nature. Most, if not all, of the adipose cell types contain their own endogenous clock. Furthermore, the relative composition of adipose tissue varies depending upon metabolic state; for instance, obesity is characterized by increased macrophage infiltration into the tissue. However, in a recent study [95] we have performed the histological analysis to assess possible macrophages contamination in human explants of adipose tissue from obese subjects. Our results showed that major changes in gene expression were exclusively a consequence of a differential transcription pattern in adipocytes. Despite these data, in most cases is the uncertainty of which cell type(s) are contributing towards the observed rhythmicity.

(c) *Other techniques*

- *Bioluminescence imaging of cell culture*
  This technique has proven to be useful for detecting protein-protein interactions, for tracking cells in vivo, and for monitoring the transcriptional and post-transcriptional regulation of specific genes with a high temporal resolution from a small amount of tissue. Recent applications have included longitudinal monitoring of tumour progression *in vivo*, and monitoring circadian rhythms with single-cell resolution [101]. In this sense, our own research group using this technique is developing the protocol to measure the circadian expression of clock genes in human adipocytes without sampling at different times of day.

A Map of Phases in Adipose Tissue

Energy metabolism and circadian systems have evolved together over millions of years to optimize internal coordination among multiple physiological and molecular processes. The adipose tissue is a metabolically active organ presenting a highly rhythmic behaviour [84]. Each cytokine has to be secreted at the right time and order in order to achieve a concerted function. To date, most published studies have been discussed as if organisms showed only one or few circadian rhythms at a time; however, circadian rhythmicity is exhibited by many variables simultaneously, raising the issue of how do the multiple rhythms relate to each other to generate a precise internal temporal order which is relevant to maintain health. Thus, an adequate temporal order in the daily pattern of genes the different cytokines and proteins implicated in adipose tissue metabolism could have important consequences not only in body fat distribution but also in the metabolic alterations associated to obesity.

Therefore, a recent study performed by our group of research has provided an overall view of the internal temporal order of circadian rhythms in human adipose tissue represented in a phase map (Fig. 3.3) [102]. The data included various genes implicated in metabolic processes such as energy intake and expenditure, insulin resistance, adipocyte differentiation, dyslipidemia, and body fat distribution, and indicated that circadian rhythmicity of the genes studied followed a predictable physiological pattern, particularly for subcutaneous depot.

Timing Interconnections Among Different Adipokines

It is well known that feeding is subject to circadian regulation [84]. Indeed, food intake is a major physiological function in animals and must be entrained to the circadian oscillations in food availability [103]. As expected, leptin, anorexigenic hormone, showed its acrophase (maximum expression) during the night (at 0200 h) in adipose tissue, coinciding with works performed in human plasma and other that demonstrate that leptin and other humoral signals are capable of communicating the nutritional state of the organism to the hypothalamic centres that control hunger and

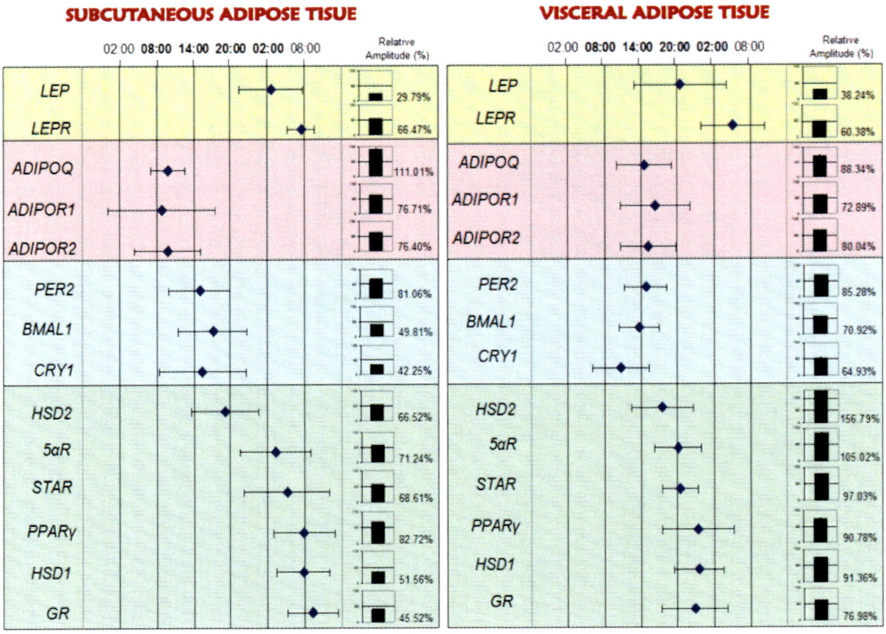

Fig. 3.3 Rhythmic expression of genes studied: (leptin and its receptor (*LEPR*), adiponectin and its receptors (*ADIPOR1* and *ADIPOR2*), clock genes (*PER2, BMAL1,* and *CRY1*) and glucocorticoid metabolism-related genes (*PPARγ, GR, HSD1, HSD2,* and *5αR*)) in human subcutaneous (**a**) and visceral adipose tissue (**b**). Adipose depots were isolated at 6-h intervals over the course of the day from adipose tissue cultures (time at 0, 6, 12, and 18 h). Results are presented relative to the lowest basal relative expression for each gene. Data of relative expression are represented as arbitrary units (AU). Data are reported as means

satiety, in a circadian-dependent manner [79, 104]. The nocturnal increase in leptin levels indicates its role as a satiety hormone, favouring fasting and nocturnal rest.

Adiponectin, the adipose tissue most abundant secreted protein, is highly implicated in glucose metabolism [105]. It has been called the fat-burning molecule because it is able to redirect fatty acids to the muscle for their oxidation. The expression of adiponectin (*ADIPOQ*) achieved its zenith (maximum) during the morning (at 1000 h) which could be implicated in the maximal withdrawal of fatty acids, and the improvement in glucose tolerance and that time [97].

Of note are the relationships between *ADIPOQ, LEPTIN,* and glucocorticoid-related genes circadian profiles and these data in human adipose tissue are consistent with previous findings in serum adiponectin and leptin variations showing out-of-phase 24-h profiles [78]. With respect to cortisol receptor (*GR*), *ADIPOQ* followed similar 24-h rhythmicity. However, although *ADIPOQ* and *GR* reached peak levels around the same time, *ADIPOQ* reached its acrophase 2 h after *GR*. These results are consistent with previous data obtained in plasma from healthy men, and highlight the tightly interactions between AT proteins [78].

PPARγ could be also related to *ADIPOQ* circadian pattern. In fact, the high expression of *PPAR*γduring the morning (0800 h), located at the beginning of the of the daily activity, is consistent with results obtained in nocturnal mammals [106] and could be influencing the further increase in *ADIPOQ* expression and the increase in insulin sensitivity during this time of the day.

Other genes studied, *glucocorticoid-related genes such as GR, and the isoenzyme 11β-hydroxysteroid dehydrogenase type 1, (HSD1),* showed their acrophase in the morning (around 0800 h). It has been described that in all species the maximum of corticosteroid rhythms occurs just before or at the onset of activity [107]. In plasma, similarly to what happened in the case performed in adipose tissue, glucocorticoids start to climb from baseline levels about 4–5 h prior to the time of waking, reaching a peak near the time of waking [40]. Over the course of the day they fall, reaching low or undetectable levels an hour or two before bedtime.

It is well known that the corticosteroid rhythm is normally tightly synchronized to the day–night and sleep–wake cycles. The antiphase relationship between leptin and glucocorticoids shown in the current study is reasonable considering that both hormones are strongly interrelated [108] and they exert opposite functions in food intake regulation. While leptin displays an anorexigenic role, glucocorticoids increase appetite (orexigenic function). In fact, in a previous work performed in plasma, leptin ultradian pulses were also inversely correlated with those of ACTH and cortisol [78].

Causation

Regarding the existence of an internal temporal order, an interesting issue concerns causation. Currently, circadian physiologists try to elucidate which genes can be driven by the circadian molecular clock and therefore can be considered as Clock Controlled Genes (CCGs). This is a very difficult question to answer. However, if we observe the clock genes circadian rhythms in the phase map, the advance of phases that suffered *BMAL1* and *CRY1* in visceral with respect to subcutaneous AT, was accompanied with a phase advance in most of the genes studied. Particularly, for *PPAR*γ and Glucocorticoid-related genes. Moreover, the phase-relationship of *BMAL1*, and *PPAR*γ is maintained in both AT depots. Previously, it has been described that *PPAR*$\gamma$is a CCG, which is activated by the positive limb, the heterodimmer CLOCK-BMAL1. Moreover, in a study performed in mPer2−/− mice the glucocorticoid rhythms disappeared suggesting that corticosteroids could be considered as CCGs [109].

Differences Between Visceral and Subcutaneous AT

Visceral AT behaved in a different way than subcutaneous for most of the genes studied. Differences were more evident for leptin and glucocorticoids-related genes, both highly related to food intake. The unexpected lower values of leptin and higher

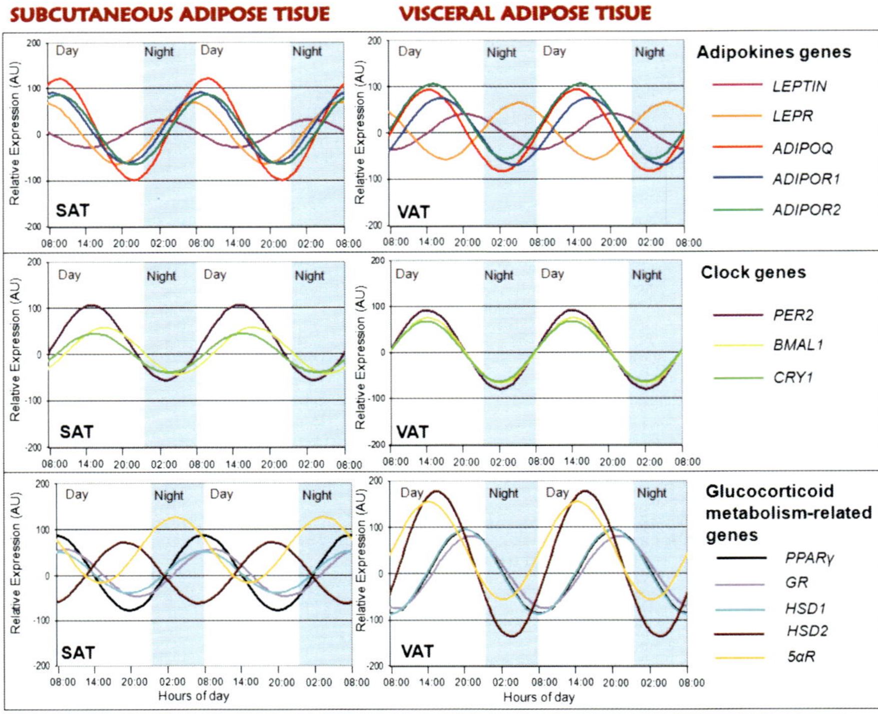

Fig. 3.4 A phase map of circadian rhythms of several genes (leptin and its receptor (*LEPR*), adiponectin and its receptors (*ADIPOR1* and *ADIPOR2*), clock genes (*PER2*, *BMAL1*, and *CRY1*), and glucocorticoid metabolism-related genes (*PPARγ*, *GR*, *HSD1*, *HSD2*, *STAR*, and *5αR*)) implicated in human adipose tissue metabolism subcutaneous adipose tissue (**a**), visceral adipose tissue (**b**). This figure shows the acrophase (time of occurrence of the best-fit maximum value) of numerous rhythms. The mean values of acrophases are plotted ± SEM

values of glucocorticoid-related genes during night hours in visceral AT could be accounting for food intake behavioural alterations already described in subjects with a predominance of visceral fat [40]. Night eaters are typically abdominal obese, show anorexia in the morning, hyperphagia in the evening and insomnia at night with frequent awakenings accompanied by food intake [110].

In subcutaneous fat, from the total genes analyzed adiponectin showed the highest circadian rhythmicity, followed by *PPARγ* and *PER2*. For instance, in visceral fat, glucocorticoid-related genes were the genes with the highest amplitude. These data reinforce the particular relevance of adiponectin in subcutaneous and glucocorticoids in visceral fat, already described in previous researches (Fig. 3.4) [105, 111, 112]. In general the relative amplitude of the genes studied was high in this study, as compared previous work carried out in different organs or tissues [113, 114].

Conclusion

For many years, adipose tissue has been considered a single stock of fat tissue. But this idea has completely changed since the last time and today the adipose tissue is defined as a true endocrine organ capable of synthesizing numerous cytokines and other factors that are involved in the metabolism. In addition, the presence of an active circadian clock in adipose tissue depots suggests that there is a temporal component to the regulation of adipose tissue function. Metabolism and maintenance of energy homeostasis require functional coordination among individual adipose depots and other metabolically active tissue sites, to insure proper nutrient/energy flux and substrate use by the organism. Therefore, further investigations of circadian rhythms in adipose tissues will provide insight into the physiology of energy homeostasis and the etiology of metabolic diseases such as obesity.

Summary Points

- Adipose tissue, until recently, considered a single stock of fat tissue, is today defined as a true endocrine organ capable of synthesizing numerous cytokines and other factors that are involved in the metabolism.
- The current scientific literature is replete with investigations providing a revolution in the study of adipose tissue biology. Many genes in adipose tissue show circadian rhythmicity.
- Many of the genes that show circadian rhythmicity in its expression in adipose tissue are involved in important metabolic processes such as energy intake and expenditure, insulin resistance, adipocyte differentiation, dyslipidemia, and body fat distribution.
- An adequate temporal order in the daily pattern of genes the different cytokines and proteins implicated in adipose tissue metabolism could have important consequences not only in body fat distribution but also in the metabolic alterations associated to obesity.

References

1. Siiteri PK (1987) Adipose tissue as a source of hormones. Am J Clin Nutr 45(1 Suppl):277–282
2. Zhang Y, Proenca R, Maffei M, Barone M, Leopold L, Friedman JM (1994) Positional cloning of the mouse obese gene and its human homologue. Nature 372:425–432
3. Friedman JM (2002) The function of leptin in nutrition, weight, and physiology. Nutr Rev 60(10(Pt 2)):S1–S14, Review
4. Zhang W, Della-Fera MA, Hartzell DL, Hausman D, Baile CA (2008) Adipose tissue gene expression profiles in ob/ob mice treated with leptin. Life Sci 83(1–2):35–42
5. Kielar D, Clark JS, Ciechanowicz A, Kurzawski G, Sulikowski T, Naruszewicz M (1998) Leptin receptor isoforms expressed in human adipose tissue. Metabolism 47(7):844–847

6. Saad MF, Riad-Gabriel MG, Khan A, Sharma A, Michael R, Jinagouda SD, Boyadjian R, Steil GM (1998) Diurnal and ultradian rhythmicity of plasma leptin: effects of gender and adiposity. J Clin Endocrinol Metab 83(2):453–459
7. Fain JN, Madan AK, Hiler ML, Cheema P, Bahouth SW (2004) Comparison of the release of adipokines by adipose tissue, adipose tissue matrix, and adipocytes from visceral and subcutaneous abdominal adipose tissues of obese humans. Endocrinology 145:2273–2282
8. Ruan H, Miles PD, Ladd CM, Ross K, Golub TR, Olefsky JM, Lodish HF (2002) Profiling gene transcription in vivo reveals adipose tissue as an immediate target of tumor necrosis factor-α: implications for insulin resistance. Diabetes 51:3176–3188
9. Ruan H, Lodish HF (2003) Insulin resistance in adipose tissue: direct and indirect effects of tumor necrosis factor. Cytokine Growth Factor Rev 14:447–455
10. Maeda K, Okubo K, Shimomura I, Funahashi T, Matsuzawa Y, Matsubara K (1996) cDNA cloning and expression of a novel adipose specific collagen-like factor, apM1 (AdiPose Most abundant Gene transcript 1). Biochem Biophys Res Commun 221:286–289
11. Scherer PE, Williams S, Fogliano M, Baldini G, Lodish HF (1995) A novel serum protein similar to C1q, produced exclusively in adipocytes. J Biol Chem 270:26746–26749
12. Nakano Y, Tobe T, Choi-Miura NH, Mazda T, Tomita M (1996) Isolation and characterization of GBP28, a novel gelatin-binding protein purified from human plasma. J Biochem (Tokyo) 120:803–812
13. Hu E, Liang P, Spiegelman BM (1996) AdipoQ is a novel adipose-specific gene dysregulated in obesity. J Biol Chem 271:10697–10703
14. Gil-Campos M, Cañete RR, Gil A (2004) Adiponectin, the missing link in insulin resistance and obesity. Clin Nutr 23:963–974
15. Yamauchi T, Kamon J, Ito Y, Tsuchida A, Yokomizo T, Kita S, Sugiyama T, Miyagishi M, Hara K, Tsunoda M, Murakami K, Ohteki T, Uchida S, Takekawa S, Waki H, Tsuno NH, Shibata Y, Terauchi Y, Froguel P, Tobe K, Koyasu S, Taira K, Kitamura T, Shimizu T, Nagai R, Kadowaki T (2003) Cloning of adiponectin receptors that mediate antidiabetic metabolic effects. Nature 423(6941):762–769
16. Rasmussen MS, Lihn AS, Pedersen SB, Bruun JM, Rasmussen M, Richelsen B (2006) Adiponectin receptors in human adipose tissue: effects of obesity, weight loss, and fat depots. Obesity (Silver Spring) 14(1):28–35
17. Steppan CM, Bailey ST, Bhat S, Brown EJ, Banerjee RR, Wright CM, Patel HR, Ahima RS, Lazar MA (2001) The hormone resistin links obesity to diabetes. Nature 409:307–312
18. Banerjee RR, Lazar MA (2003) Resistin: molecular history and prognosis. J Mol Med 81:218–226
19. Kim KH, Lee K, Moon YS, Sul HS (2001) A cysteine-rich adipose tissue-specific secretory factor inhibits adipocyte differentiation. J Biol Chem 276(14):11252–11256
20. Adeghate E (2004) An update on the biology and physiology of resistin. Cell Mol Life Sci 61(19–20):2485–2496
21. Steppan CM, Lazar MA (2004) The current biology of resistin. J Intern Med 255:439–447
22. Dyck DJ, Heigenhauser GJ, Bruce CR (2006) The role of adipokines as regulators of skeletal muscle fatty acid metabolism and insulin sensitivity. Acta Physiol 186:5–16
23. McTernan PG, Kusminski CM, Kumar S (2006) Resistin. Curr Opin Lipidol 17(2):170–175
24. Fernandez-Real JM, Ricart W (2003) Insulin resistance and chronic cardiovascular inflammatory syndrome. Endocr Rev 24:278–301
25. Kershaw EE, Flier JS (2004) Adipose tissue as an endocrine organ. J Clin Endocrinol Metab 89(6):2548–2556
26. Cianflone K, Xia Z, Chen LY (2003) Critical review of acylation-stimulating protein physiology in humans and rodents. Biochim Biophys Acta 1609:127–143
27. Flier JS, Cook KS, Usher P, Spiegelman BM (1987) Severely impaired adipsin expression in genetic and acquired obesity. Science 237:405–408
28. Fukuhara A, Matsuda M, Nishizawa M, Segawa K, Tanaka M, Kishimoto K, Matsuki Y, Murakami M, Ichisaka T, Murakami H, Watanabe E, Takagi T, Akiyoshi M, Ohtsubo T, Kihara S, Yamashita S, Makishima M, Funahashi T, Yamanaka S, Hiramatsu R, Matsuzawa Y,

Shimomura I (2005) Visfatin: a protein secreted by visceral fat that mimics the effects of insulin. Science 307:426–430
29. Revollo JR (2007) Nampt/PBEF/Visfatin regulates insulin secretion in beta cells as a systemic NAD biosynthetic enzyme. Cell Metab 6:363–375
30. Frühbeck G (2004) The adipose tissue as a source of vasoactive factors. Curr Med Chem Cardiovasc Hematol Agents 2(3):197–208
31. Maeda K, Okubo K, Shimomura I, Mizuno K, Matsuzawa Y, Matsubara K (1997) Analysis of an expression profile of genes in the human adipose tissue. Gene 190:227–235
32. Stewart PM, Petersenn S (2009) Rationale for treatment and therapeutic options in Cushing's disease. Best Pract Res Clin Endocrinol Metab 23(Suppl 1):S15–S22
33. McDonough AK, Curtis JR, Saag KG (2008) The epidemiology of glucocorticoid-associated adverse events. Curr Opin Rheumatol 20:131–137
34. Friedman TC, Mastorakos G, Newman TD, Mullen NM, Horton EG, Costello R, Papadopoulos NM, Chrousos GP (1996) Carbohydrate and lipid metabolism in endogenous hypercortisolism: shared features with metabolic syndrome X and NIDDM. Endocr J 43(6):645–655
35. Bujalska IJ, Kumar S, Stewart PM (1997) Does central obesity reflect "Cushing's disease of the omentum"? Lancet 349:1210–1213
36. Andrew R, Phillips DI, Walker BR (1998) Obesity and gender influence cortisol secretion and metabolism in man. J Clin Endocrinol Metab 83:1806–1809
37. Strain GW, Zumoff B, Strain JJ, Levin J, Fukushima DK (1980) Cortisol production in obesity. Metabolism 29:980–985
38. Westerbacka J, Yki-Järvinen H, Vehkavaara S, Häkkinen AM, Andrew R, Wake DJ, Seckl JR, Walker BR (2003) Body fat distribution and cortisol metabolism in healthy men: enhanced 5beta-reductase and lower cortisol/cortisone metabolite ratios in men with fatty liver. J Clin Endocrinol Metab 88:4924–4931
39. Fraser R, InhGram MC, Anderson NH, Morrison C, Davies E, Connell JM (1999) Cortisol effects on body mass, blood pressure, and cholesterol in the general population. Hypertension 33:1364–1368
40. García-Prieto MD, Tébar FJ, Nicolás F, Larqué E, Zamora S, Garaulet M (2007) Cortisol secretary pattern and glucocorticoid feedback sensitivity in women from a Mediterranean area: relationship with anthropometric characteristics, dietary intake and plasma fatty acid profile. Clin Endocrinol (Oxf) 66:185–191
41. Pezzi V, Mathis JM, Rainey WE, Carr BR (2003) Profiling transcript levels for steroidogenic enzymes in fetal tissues. J Steroid Biochem Mol Biol 87:181–189
42. Schinner S, Willenberg HS, Krause D, Schott M, Lamounier-Zepter V, Krug AW, Ehrhart-Bornstein M, Bornstein SR, Scherbaum WA (2007) Adipocyte-derived products induce the transcription of the StAR promoter and stimulate aldosterone and cortisol secretion from adrenocortical cells through the Wnt-signaling pathway. Int J Obes (Lond) 31:864–870
43. Livingstone DE, Jones GC, Smith K, Jamieson PM, Andrew R, Kenyon CJ, Walker BR (2003) Understanding the role of glucocorticoids in obesity: tissue-specific alterations of corticosterone metabolism in obese Zucker rats. Endocrinology 141:560–563
44. Gerhold DL, Liu F, Jiang G, Li Z, Xu J, Lu M, Sachs JR, Bagchi A, Fridman A, Holder DJ, Doebber TW, Berger J, Elbrecht A, Moller DE, Zhang BB (2002) Gene expression profile of adipocyte differentiation and its regulation by peroxisome proliferator-activated receptor-gamma agonists. Endocrinology 143(6):2106–2118
45. Tomlinson JW, Moore J, Cooper MS, Bujalska I, Shahmanesh M, Burt C, Strain A, Hewison M, Stewart PM (2001) Regulation of expression of 11beta-hydroxysteroid dehydrogenase type 1 in adipose tissue: tissue-specific induction by cytokines. Endocrinology 142: 1982–1989
46. Berger J, Tanen M, Elbrecht A, Hermanowski-Vosatka A, Moller DE, Wright SD, Thieringer R (2001) Peroxisome proliferator-activated receptor-gamma ligands inhibit adipocyte 11beta -hydroxysteroid dehydrogenase type 1 expression and activity. J Biol Chem 276(16): 12629–12635
47. Fawcett DW (1952) A comparison of the histological organization and cytochemical reactions of brown and white adipose tissues. J Morphol 90:363–405

48. Napolitano L (1963) The differentiation of white adipose tissue: an electron microscopic study. J Cell Biol 18:663–673
49. Afzelius BA (1970) Brown adipose tissue: its gross anatomy, histology, and cytology. In: Lindberg O (ed) Brown adipose tissue. Elsevier, Amsterdam, pp 1–31
50. Cannon B, Nedergaard J (2004) Brown adipose tissue: function and physiological significance. Physiol Rev 84:277–359
51. Garaulet M, Hernandez-Morante JJ, Lujan J, Tebar FJ, Zamora S (2006) Relationship between fat cell size and number and fatty acid composition in adipose tissue from different fat depots in overweight/obese humans. Int J Obes (Lond) 30(6):899–905
52. Gil A, Olza J, Gil-Campos M, Gomez-Llorente C, Aguilera CM (2011) Is adipose tissue metabolically different at different sites? Int J Pediatr Obes 6(Suppl 1):13–20, Review
53. Rebuffé-Scrive M, Lundholm K, Björntorp P (1985) Glucocorticoid hormone binding to human adipose tissue. Eur J Clin Invest 15:267–271
54. Freedland ES (2004) Role of critical visceral adipose tissue threshold in metabolic syndrome: implications for controlling dietary carbohydrates: a review. Nutr Metab 1:12
55. Pedersen SB, Hansen PS, Lund S, Andersen PH, Odgaard A, Richelsen B (1996) Identification of oestrogen receptors and oestrogen receptor mRNA in human adipose tissue. Eur J Clin Invest 26:259
56. Imbeault P, Couillard C, Tremblay A, Després J-P, Mauriège P (2000) Reduced alpha [2]-adrenergic sensitivity of subcutaneous abdominal adipocytes as a modulator of fasting and post-prandial triglyceride levels in men. J Lipid Res 41:1367
57. Björntorp P (2000) Metabolic difference between visceral fat and subcutaneous abdominal fat. Diabetes Metab 26(Suppl 3):10–12, Review. French
58. Amer P (1995) Differences in lipolysis between human subcutaneous and omental adipose tissues. Ann Med 27:435–438
59. Arner P (1997) Obesity and the adipocyte. Regional adipocity in man. J Endocrinol 155:191–192
60. Hamdy O, Porramatikul S, Al-Ozairi E (2006) Metabolic obesity: the paradox between visceral and subcutaneous fat. Curr Diabetes Rev 2(4):367–373
61. Prasad H, Ryan DA, Celzo MF, Stapleton D (2012) Metabolic syndrome: definition and therapeutic implications. Postgrad Med 124(1):21–30
62. Oh TH, Byeon JS, Myung SJ, Yang SK, Choi KS, Chung JW, Kim B, Lee D, Byun JH, Jang SJ, Kim JH (2008) Visceral obesity as a risk factor for colorectal neoplasm. J Gastroenterol Hepatol 23:411–417
63. Von Hafe P, Pina F, Pérez A, Tavares M, Barros H (2004) Visceral fat accumulation as a risk factor for prostate cancer. Obes Res 12:1930–1935
64. Garaulet M, Madrid JA (2009) Chronobiology, genetics and metabolic syndrome. Curr Opin Lipidol 20(2):127–134, Review
65. Dunlap JC (1999) Molecular bases for circadian clocks. Cell 96(2):271–290
66. Lee C, Etchegaray JP, Cagampang FR, Loudon AS, Reppert SM (2001) Posttranslational mechanisms regulate the mammalian circadian clock. Cell 107(7):855–867
67. Pevet P, Challet E (2011) Melatonin: both master clock output and internal time-giver in the circadian clocks network. J Physiol Paris 105(4–6):170–182
68. Balsalobre A, Damiola F, Schibler U (1998) A serum shock induces circadian gene expression in mammalian tissue culture cells. Cell 93(6):929–937
69. Arble DM, Ramsey KM, Bass J, Turek FW (2010) Circadian disruption and metabolic disease: findings from animal models. Best Pract Res Clin Endocrinol Metab 24(5):785–800
70. Aoyagi T, Shimba S, Tezuka M (2005) Characteristics of circadian gene expressions in mice white adipose tissue and 3T3-L1 adipocytes. J Health Sci 51:21–32
71. Loboda A, Kraft WK, Fine B, Joseph J, Nebozhyn M, Zhang C, He Y, Yang X, Wright C, Morris M, Chalikonda I, Ferguson M, Emilsson V, Leonardson A, Lamb J, Dai H, Schadt E, Greenberg HE, Lum PY (2009) Diurnal variation of the human adipose transcriptome and the link to metabolic disease. BMC Med Genomics 2:7
72. Ptitsyn AA, Zvonic S, Conrad SA, Scott LK, Mynatt RL, Gimble JM (2006) Circadian clocks are resounding in peripheral tissues. PLoS Comput Biol 2(3):e16

73. Bozek K, Relógio A, Kielbasa SM, Heine M, Dame C, Kramer A, Herzel H (2009) Regulation of clock-controlled genes in mammals. PLoS One 4(3):e4882
74. Dallmann R, Viola AU, Tarokh L, Cajochen C, Brown SA (2012) The human circadian metabolome. Proc Natl Acad Sci U S A 109(7):2625–2629
75. Le Magnen J (1988) Lipogenesis, lipolysis and feeding rhythms. Ann Endocrinol Paris 49(2):98–104, French
76. Gimble JM, Floyd ZE (2009) Fat circadian biology. J Appl Physiol 107(5):1629–1637
77. Sinha MK, Ohanneslan JP, Heiman ML, Kriauclunas A, Stephens TW, Magosin S, Marco C, Caro JF (1996) Nocturnal rise of leptin in lean, obese, and non-insulin-dependent diabetes mellitus subjects. J Clin Invest 97:1344–1347
78. Gavrila A, Peng CK, Chan JL, Mietus JE, Goldberger AL, Mantzoros CS (2003) Diurnal and ultradian dynamics of serum adiponectin in healthy men: comparison with leptin, circulating soluble leptin receptor, and cortisol patterns. J Clin Endocrinol Metab 88:2838–2843
79. Langendonk JG, Pijl H, Toornvliet AC, Burggraaf J, Frölich M, Schoemaker RC, Doornbos J, Cohen AF, Meinders AE (1998) Circadian rhythm of plasma leptin levels in upper and lower body obese women: influence of body fat distribution and weight loss. J Clin Endocrinol Metab 83(5):1706–1712
80. Kalsbeek A, Fliers E, Romijn JA, La Fleur SE, Wortel J, Bakker O, Endert E, Buijs RM (2001) The suprachiasmatic nucleus generates the diurnal changes in plasma leptin levels. Endocrinology 142:2677–2685
81. Zvonic S, Ptitsyn AA, Conrad SA, Scott LK, Floyd ZE, Kilroy G, Wu X, Goh BC, Mynatt RL, Gimble JM (2006) Characterization of peripheral circadian clocks in adipose tissues. Diabetes 55(4):962–970
82. Ando H, Yanagihara H, Hayashi Y, Obi Y, Tsuruoka S, Takamura T, Kaneko S, Fujimura A (2005) Rhythmic messenger ribonucleic acid expression of clock genes and adipocytokines in mouse visceral adipose tissue. Endocrinology 146(12):5631–5636
83. Suzuki M, Shimomura Y, Satoh Y (1983) Diurnal changes in lipolytic activity of isolated fat cells and their increased responsiveness to epinephrine and theophylline with meal feeding in rats. J Nutr Sci Vitaminol (Tokyo) 29(4):399–411
84. Garaulet M, Madrid JA (2010) Chronobiological aspects of nutrition, metabolic syndrome and obesity. Adv Drug Deliv Rev 62:967–978
85. Chawla A, Lazar MA (1993) Induction of Rev-ErbAa, an orphan receptor encoded on the opposite strand of the a-thyroid hormone receptor gene, during adipocyte differentiation. J Biol Chem 268:16265–16269
86. Shimba S, Ishii N, Ohno T, Watabe Y, Hayashi M, Wada T, Aoyagi T, Tezuka M (2005) Brain and muscle Arnt-like protein-1 (BMAL1), a component of the molecular clock, regulates adipogenesis. Proc Natl Acad Sci U S A 102:12071–12076
87. Benito J, Hoxha V, Lama C, Lazareva AA, Ferveur JF, Hardin PE, Dauwalder B (2010) The circadian output gene takeout is regulated by Pdp1epsilon. Proc Natl Acad Sci U S A 107:2544–2549
88. Sarov-Blat L, So WV, Liu L, Rosbash M (2000) The Drosophila takeout gene is a novel molecular link between circadian rhythms and feeding behavior. Cell 101:647–656
89. Gómez-Abellán P, Hernández-Morante JJ, Luján JA, Madrid JA, Garaulet M (2008) Clock genes are implicated in the human metabolic syndrome. Int J Obes (Lond) 32:121–128
90. Gómez-Abellán P, Madrid JA, Luján JA, Frutos MD, González R, Martínez-Augustín O, de Medina FS, Ordovás JM, Garaulet M (2012) Sexual dimorphism in clock genes expression in human adipose tissue. Obes Surg 22(1):105–112
91. Zanquetta MM, Correa-Giannella ML, Giannella-Neto D, Alonso PA, Guimarães LM, Meyer A, Villares SM (2012) Expression of clock genes in human subcutaneous and visceral adipose tissues. Chronobiol Int 29(3):252–260
92. Gómez-Santos C, Gómez-Abellán P, Madrid JA, Hernández-Morante JJ, Lujan JA, Ordovas JM, Garaulet M (2009) Circadian rhythm of clock genes in human adipose explants. Obesity (Silver Spring) 17(8):1481–1485

93. Gómez-Abellán P, Madrid JA, Diez-Noguera A, Luján JA, Ordovás JM, Garaulet M. Resetting of circadian time of CLOCK in human adipose tissue by glucocorticoid signaling. PLoS One. 2012 (in press).
94. Otway DT, Mäntele S, Bretschneider S, Wright J, Trayhurn P, Skene DJ, Robertson MD, Johnston JD (2011) Rhythmic diurnal gene expression in human adipose tissue from individuals who are lean, overweight, and type 2 diabetic. Diabetes 60(5):1577–1581
95. Gomez-Santos C, Hernandez-Morante JJ, Margareto J, Larrarte E, Formiguera X, Martínez CM, Garaulet M (2011) Profile of adipose tissue gene expression in premenopausal and postmenopausal women: site-specific differences. Menopause 18(6):675–684
96. Wu X, Zvonic S, Floyd ZE, Kilroy G, Goh BC, Hernandez TL, Eckel RH, Mynatt RL, Gimble JM (2007) Induction of circadian gene expression in human subcutaneous adipose-derived stem cells. Obesity (Silver Spring) 15(11):2560–2570
97. Gómez-Abellán P, Gómez-Santos C, Madrid JA, Milagro FI, Campion J, Martínez JA, Ordovás JM, Garaulet M (2010) Circadian expression of adiponectin and its receptors in human adipose tissue. Endocrinology 151(1):115–122
98. Gómez Abellán P, Gómez Santos C, Madrid JA, Milagro FI, Campion J, Martínez JA, Luján JA, Ordovás JM, Garaulet M (2011) Site-specific circadian expression of leptin and its receptor in human adipose tissue. Nutr Hosp 26(6):1394–1401
99. Hernandez-Morante JJ, Gomez-Santos C, Milagro F, Campión J, Martínez JA, Zamora S, Garaulet M (2009) Expression of cortisol metabolism-related genes shows circadian rhythmic patterns in human adipose tissue. Int J Obes (Lond) 33(4):473–480
100. Khan S, Minihane AM, Talmud PJ, Wright JW, Murphy MC, Williams CM, Griffin BA (2002) Dietary long-chain n-3 PUFAs increase LPL gene expression in adipose tissue of subjects with an atherogenic lipoprotein phenotype. J Lipid Res 43(6):979–985
101. Welsh DK, Kay SA (2005) Bioluminescence imaging in living organisms. Curr Opin Biotechnol 16(1):73–78
102. Garaulet M, Ordovás JM, Gómez-Abellán P, Martínez JA, Madrid JA (2011) An approximation to the temporal order in endogenous circadian rhythms of genes implicated in human adipose tissue metabolism. J Cell Physiol 226(8):2075–2080
103. Dietrich MO, Horvath TL (2009) Feeding signals and brain circuitry. Eur J Neurosci 30:1688–1696
104. Kalra SP, Bagnasco M, Otukonyong EE, Dube MG, Kalra PS (2003) Rhythmic, reciprocal ghrelin and leptin signaling: new insight in the development of obesity. Regul Pept 111:1–11
105. Garaulet M, Hernández-Morante JJ, de Heredia FP, Tébar FJ (2007) Adiponectin, the controversial hormone. Public Health Nutr 10:1145–1150
106. Yang X, Downes M, Yu RT, Bookout AL, He W, Straume M, Mangelsdorf DJ, Evans RM (2006) Nuclear receptor expression links the circadian clock to metabolism. Cell 126:801–810
107. Peterson RE (1957) Plasma corticosterone and hydrocortisone levels in man. J Clin Endocrinol Metab 17:1150–1157
108. Henry BA, Clarke IJ (2008) Adipose tissue hormones and the regulation of food intake. J Neuroendocrinol 20:842–849
109. Yang S, Liu A, Weidenhammer A, Cooksey RC, McClain D, Kim MK, Aguilera G, Abel ED, Chung JH (2009) The role of mPer2 clock gene in glucocorticoid and feeding rhythms. Endocrinology 150:2153–2160
110. Qin LQ, Li J, Wang Y, Wang J, Xu JY, Kaneko T (2003) The effects of nocturnal life on endocrine circadian patterns in healthy adults. Life Sci 73:2467–2475
111. Marin P, Darin N, Amemiya T, Andersson B, Jern S, Björntorp P (1992) Cortisol secretion in relation to body fat distribution in obese premenopausal women. Metabolism 41:882–886
112. Björntorp P (1991) Adipose tissue distribution and function. Int J Obes 15:67–81

113. Hermida RC, Ayala DE, Calvo C, López JE, Fernández JR, Mojón A, Domínguez MJ, Covelo M (2003) Administration-time dependent effects of acetyl-salicylic acid on blood pressure in patients with mild essential hypertension. Med Clin (Barc) 120:686–692
114. Hermida RC, Calvo C, Ayala DE, López JE, Fernández JR, Mojón A, Domínguez MJ, Covelo M (2003) Seasonal variation in plasma fibrinogen in dipper and non-dipper patients with mildmoderate essential hypertension. Med Clin (Barc) 121:6–11
115. Johnston JD (2012) Adipose circadian rhythms: translating cellular and animal studies to human physiology. Mol Cell Endocrinol 349(1):45–50

Chapter 4
Processes Underlying Chronodisruption and Their Proposed Association with Illness

Russel J. Reiter, Xiaoyan Liu, Lucien C. Manchester, Sergio A. Rosales-Corral, Dun-Xian Tan, and Juan Antonio Madrid Pérez

Abstract Regularly alternating periods of light and darkness, such as normally occur with the rising and the setting of the sun, are essential for the maintenance of undisturbed circadian rhythms in all organisms including humans. The light–dark environment, as detected by specialized photoreceptors in the retinas, impacts the endogenous circadian clock in the anterior hypothalamus, the suprachiasmatic nuclei. These nuclei, via both neural and humoral signals, communicate with cells throughout the organism to establish regular circadian rhythms. The introduction of artificial sources of light roughly 150 years ago has significantly undermined the naturally occurring light–dark environment and, likewise, has disturbed circadian rhythms since light is now available at unusual times, i.e., at night. Light at night is known to cause circadian disruption and melatonin suppression. Of many potentially pathophysiological consequences of these artificial light-mediated changes, female breast cancer has become of major interest. Additionally, however, there is currently data suggesting that not only breast cancer, but cancer in general, cardiovascular diseases, insomnia, metabolic syndrome, and affective and cognitive disorders may be aggravated by the increased exposure to light at night, which is inevitable in well-developed societies that have undergone extensive electrification.

R.J. Reiter, Ph. D. (✉) • X. Liu • L. C. Manchester, Ph.D. • S.A. Rosales-Corral, MD
• D.-X. Tan, MD, Ph. D.
Department of Cellular and Structural Biology, University of Texas Health Science Center at San Antonio, 7703 Floyd Curl Drive, San Antonio, TX 78229, USA
e-mail: reiter@uthscsa.edu

J.A. Madrid Pérez, Ph.D.
Chronobiology Laboratory, Department of Physiology, Faculty of Biology, University of Murcia, Murcia, Spain

M. Garaulet and J.M. Ordovás (eds.), *Chronobiology and Obesity*,
DOI 10.1007/978-1-4614-5082-5_4,

Abbreviations

SCN	Suprachiasmatic nucleus
CD	Chronodisruption
LD	Light–dark
LAN	Light at night
ACTH	Adrenocorticotropic hormone
BMAL1 or ARNTL or MOP3	Aryl hydrocarbon receptor nuclear translocator-like
PER2	Period homolog 2 (Drosophila)
CLOCK	Circadian locomotor output cycles kaput
MetS	Metabolic syndrome
WT	Wrist temperature
BP	Blood pressure

Introduction

Clearly, circadian rhythms are so commonplace in animal and human physiology that, a priori, we must assume they are relevant to optimal function. Indeed, this fact is becoming progressively more apparent as scientists examine 24-h variations in the physiology of organisms, organs, cells and cellular organelles [1–3]. With an estimated 10–20 % of the genes in each cell under control of the central oscillator, the suprachiasmatic nuclei (SCN) in the anterior diencephalon, it is easy to envisage that disturbances of these clock mechanisms, particularly when repeated or chronic, may well lead to pathologies.

While the SCN exhibits an intrinsic rhythm of approximately 24 h, corresponding roughly to the normal environment light–dark cycle, it is not precisely of 24-h duration. Thus, light information detected by specialized ganglion cells (the intrinsic photoreceptive ganglion cells, i.e., *ip*RGC) [4–6] in the retinas and the transfer of this information via the retinohypothalamic tract to the SCN, serves as an important and critical Zeitgeber for the central oscillator. This, in turn, influences the circadian physiology of all peripheral oscillators. Thus, whereas cells in organs throughout the body are incapable of direct photoreception, they do receive information as to whether it is day or night from the central clock and they adjust their functions accordingly.

The means by which the SCN conveys circadian information to the peripheral cellular oscillators includes both neural and humoral messages. One important humoral message is the melatonin signal from the pineal gland [7]. Due to information detected by the *ip*RGC of the retina and relayed through the SCN, the pinealocytes are apprised of the prevailing light–dark environment. During the day messages from the *ip*RGC render the SCN dormant relative to its ability to activate the pineal gland. During darkness, the SCN, via a circuitous neural route that involves portions of the central and peripheral sympathetic nervous system, stimulates the pinealocytes to synthesize and quickly release the chemical mediator, melatonin [7, 8]. This agent then circulates throughout the body and informs, perhaps every cell, of the prevailing photoperiodic environment so that functional adjustments can be made [9, 10].

Considering that the circadian system is composed by multiple body clocks, it is important to note that, aside from light–dark influences to the SCN, other peripheral oscillators are very sensitive to non-photic synchronizers such as feeding time, scheduled exercise, sleep time and social contacts. This complexity, together with the weakness of exposure to synchronizers in developed countries, makes the circadian system prone to suffer from chronodisruption (CD).

What Causes Chronodisruption?

Circadian disruption or CD is defined as a relevant disturbance of the internal temporal order of physiological and behavioral circadian rhythms. It is also a breakdown of the normal phase relationship between the internal circadian rhythms and 24-h environmental cycles. In our modern society, CD may be common in several conditions such as jet lag, shift work, light at night, or social jet lag [11]. In addition clock gene polymorphisms and aging may have also chronodisruptive effects. Thus, CD can be induced by any impairment of the inputs, oscillators and outputs (Fig. 4.1).

Inadequate Inputs

Importance of a Regular Light–Dark Cycle

Since the light–dark (LD) cycle regulates the 24-h melatonin synthesis/secretion rhythm, it is obvious that perturbations of this regularly recurring cycle would likewise alter the melatonin message leading to misinformation being sent to the peripheral cellular clocks. Throughout eons of evolution, disturbances of the LD environment did not occur since the SCN was regulated and the melatonin rhythm was determined exclusively by the rising and the setting of the sun. Only after the invention of the artificial light source by Thomas Alva Edison in 1879 were humans capable of imposing the LD cycle they desired. This was a major turning point for circadian biology since it allowed humans to inadvertently alter an important basic system, the circadian system, and to also perturb the melatonin rhythm. These perturbations in the melatonin rhythm could not be ignored by the peripheral cellular oscillators and they then were required to make unexpected adjustments to their physiology at unusual times. Not surprisingly, these atypical changes of a very basic

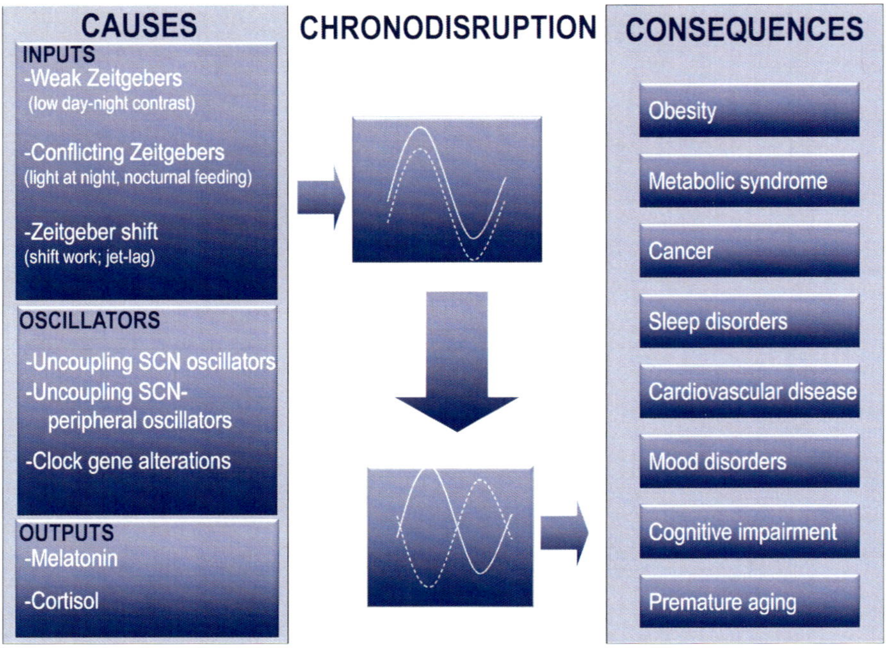

Fig. 4.1 Causes and consequences of chronodisruption. Circadian disruption is the result of an abnormal phase relationship between the rhythms regulated by endogenous oscillators (*solid line*) and activity-controlled physiological processes (*dotted line*). Chronodisruption can be induced by factors related to the following: (1) Impairment of the inputs to the circadian pacemaker: low contrast between day and night synchronizing agents (continuous light, frequent snacking, low levels of physical exercise, etc.); exposure to zeitgebers of different periods or unusual phasing (i.e., light at night, nocturnal feeding, nocturnal physical exercise) or by frequent shifts in the time provided by zeitgebers (i.e., jet lag, shift work). (2) Circadian oscillators: the uncoupling between different subpopulations of oscillators inside the SCN caused by aging or clock gene alterations and the uncoupling between central pacemaker and peripheral oscillators also result in chronodisruption. (3) Outputs: the suppression of nocturnal melatonin and the loss of cortisol rhythm are also chronodisrupters. CD is related to the increasing risk of developing certain diseases and with the impairment of preexisting pathologies

rhythm, which had been unvarying throughout evolution, would be expected to cause abnormal physiology and perhaps pathologies [2, 12, 13].

Already several decades ago it was shown that the exposure of animals or humans to light at night (LAN) depresses elevated nighttime melatonin levels, i.e., it changes the melatonin message [14–18]. Since these early discoveries in this field, numerous subsequent investigations have more precisely defined the mechanisms by which LAN inhibits the synthesis and discharge of melatonin, thereby destroying its regular circulating rhythm [6, 7].

LAN has a number of unnatural effects on the circadian system and on pineal melatonin production. In the well-developed countries of the world where electricity is readily available, it is usual to turn on manufactured lights at the time the sunsets. Likewise, when humans awaken in the morning and it is still dark, their first activity is to turn on a light. What this obviously means is that humans are no longer allowing the natural LD cycle to regulate their SCN and the production of melatonin; rather they unwittingly manipulate these processes with the misuse of artificial light.

In the situation summarized above, where light is extended into the normal period of darkness and light exposure also occurs in the morning before sunrise, the SCN must adjust its signaling processes accordingly; one result of this imposed artificial light is a shortening of the nightly duration of melatonin production by the pineal gland [19]. Thus, in highly industrialized societies, humans have become relatively melatonin deficient since they truncate the normal period of darkness and, thereby, the duration of elevated melatonin.

In general, the only time humans are in darkness is when they sleep. Thus, depending on their nightly sleep duration, which is also becoming progressively shorter in developed societies [20, 21], the total amount of melatonin produced may be reduced by 50 % or more relative to what would have been generated had they been exposed to the natural LD cycle. Given the multiple beneficial actions of melatonin [22–25], a 50 % loss of the total amount of melatonin nightly would be expected to have consequences, including negative effects, i.e., pathologies, [26–28].

A second danger of the availability of manufactured light is the ability of humans to acutely expose themselves to light during the normal dark period. During the night, the *ip*RGC detect the light that is actively imposed and they signal the SCN that it is "day," when in fact it is night. The SCN, in turn, communicates with organs throughout the organism, including the pineal gland, and provides them with misinformation. Even though the information received is inappropriate for the time, molecular and physiological adjustments are made by peripheral organs consistent with the normal daytime period. Thus, the physiology of these cells is no longer in synchrony with the normal LD cycle. This incoordination results in chronodisruption [10], a situation which could certainly lead to pathophysiologies [2, 13, 15, 29].

One conspicuous result of acute bright light exposure at night is a rapid plummeting of circulating melatonin levels [14, 17]. This abnormal drop in blood melatonin is also "read" by numerous cells throughout the body and, since melatonin values are normally low only during the day, again the peripheral organs are misled as to time. Such severe perturbations and dyssynchrony between organ physiology and the prevailing LD environment would not surprisingly be expected to cause functional casualties.

There is one other hazard associated with the massive electrification that has occurred in many places in the world. In cities especially, LAN has become unavoidable because of light pollution. Photographs of America, Europe, Japan, etc., taken from outer space at night reveals the degree of light pollution that has taken place in recent decades and predictably will become progressively worse in the years ahead. What this means is that it is becoming more difficult for humans to avoid LAN and, as a result, we may be becoming progressively more chronodisrupted. Moreover, in the sleeping environment it is sometimes difficult to evade transpass light, i.e., light that is out of the control of the person being exposed. The degree to which light pollution, transpass light, etc., impacts circadian physiology and pineal melatonin production has not been well defined, but deserves close scrutiny. Finally, parents should be discouraged from permitting their children to sleep in a lit room because they are intimidated by darkness. In the case of children, LAN may be particularly harmful since this is the age at which their circadian system is maturing. Presumably, any pathologies that may develop as a result of chronic chronodisruption may appear years or even decades after the disturbances have occurred.

Consequences of disturbing the light–dark cycle, as already mentioned, the extension of the light phase into the normal period of darkness and likewise truncating the night by artificial light exposure in the morning, limits the total amount of melatonin that would normally be produced in a seasonally appropriate LD cycle. While these maneuvers reduce the total quantity of melatonin produced, they do not eradicate its basic circadian rhythm [19]. The reduction in the total amount of melatonin produced during a 24-h period by itself may, however, be consequential in terms of at least one pathophysiological state, i.e., cancer [23, 30–32].

Sleep Deprivation

In addition to limiting the quantity of melatonin generated nightly, the excessive use of light in the evening and in the morning is usually associated with abbreviated nightly sleep periods. Adequate sleep is also essential for optimal health [33]. The daily sleep interval is becoming shorter in a number of countries. For example, in the USA the nightly sleep period has been dropping over the last half century. In the 1960s, the duration of sleep by US citizens was on the order of 8 h; the current best measures indicate that this is currently about 6.5 h [34]. Since sleep duration is usually associated with dark exposure, there has been a corresponding drop in melatonin production in the last 50 years as well. In addition to the obvious reduction in work efficiency that accompanies sleepiness, the health consequences of insufficient rest combined with reduced melatonin are likely also substantial [15, 29].

Virtually all studies that have examined a possible association between nightly sleep duration and cancer, most often female breast cancer, have found a relationship, i.e., short sleep correlated with elevated cancer incidence [35–37]. Each of these workers also noted that reduced sleep likely also meant limited elevated melatonin levels every night. Considering the multiple means by which this indole inhibits tumors, its reduction would likely be a contributing factor to the higher breast cancer incidence [30, 38, 39].

Feeding Time

Rhythmic feeding appears to be the major synchronizer for peripheral oscillators. Thus, unusual feeding time can produce CD by inducing internal desynchronization through decoupling of peripheral oscillators from the SCN. For example, it has been described that clock gene expression in the liver can synchronize to scheduled feeding in 2 days while SCN remains locked to LD cycle [40]. This differential synchronization induced by abnormal feeding habits could produce unhealthy consequences also in humans. Thus, when nocturnal (characterized by late awakening, omitting breakfast and late dinner) and diurnal (early awakening and early dinner) lifestyles were compared among healthy young people, glucose tolerance and insulin response were found to be well regulated in the diurnal group, while sustained hyperglycemia and hypoglycemia in the morning were observed in the case of nocturnal lifestyle [41] In addition, plasma levels of melatonin and leptin were reduced during the night in the nocturnal lifestyle group.

Different laboratory experiments are inducing internal desynchrony in human by forced desynchronization protocols placing subjects with sleep–wake and feeding rhythms on a 28-h daily routine for several days [42]. Interestingly, under such artificial conditions, body temperature, which is under the SCN control, remains near 24 h cycle, whereas rhythmicity in different metabolic hormones such as leptin and insulin, adhered to the imposed 28-h behavioral cycle of sleep and food intake. This misalignment induces a suppression of plasma leptin, increase in glycemia, and hypertension. Therefore, unusual feeding times are likely an overlooked risk factor to the health in modern societies.

Impaired Pacemakers

Jet lag and rotating shift-work are two well documented factors inducing CD. These two factors share a common mechanism in the CD generation involving the differential rates of synchronization of biological variables. This may be the result of the different contribution of SCN and peripheral oscillators to the generation of biological rhythms in different variables. For example, following a 6-h phase delay, the acrophase of ACTH and cortisol rhythms need up to 7 days to resynchronize, while it takes only 3 days in the case of sleep–wake cycle. Thus, during these days each function shifts at a different rate and the organism is suboptimally organized to efficiently accomplish its functions [11]. This condition is even more serious in the case of shift-workers because its chronic character.

CD can also be produced by the impairment of the molecular machinery of the circadian clock; however, discerning the relative influence of disrupted circadian rhythms induced by clock gene alteration from the potential pleiotropic effects of core clock gene inducing pathological process will therefore be a challenge [43]. Alterations in some clock genes have been associated with specific diseases such as premature aging, cancer, and obesity among others.

Mice without Bmal1 gene are arrhythmic and are subjected to premature aging and the mean lifespan of these animals is of 37 weeks as compared with 120 weeks for wild-type animals. In addition, Bmal1-knockout mice exhibit sarcopenia, vision problems, and altered lymphocytopenia [44].

Mutations in Per2 gene, another core clock gene, increases the susceptibility of mice to develop spontaneous and irradiation-induced tumors, through the stimulation of the protooncogene c-myc and repression of the oncostatic p53 gene [45]. Another well known example of clock gene induced pathology is obesity and metabolic syndrome associated with impaired Clock gene expression. CLOCK protein is a key factor involved in the synchronization of metabolic processes with the environment and in the control of mammalian energy balance. As expected, Clock-mutant mice show reduced or abolished rhythms in food intake and metabolic rate, but, in addition these animals are obese, exhibit adipocyte hypertrophy, hepatic steatosis, and alteration in leptin blood levels [29].

Outputs

The third element which can cause CD are the alteration of the outputs of the central pacemaker which act as internal coupling synchronizing signal to maintain the internal temporal order among different rhythmic functions. Impairment of the melatonin rhythm is the best known output factor mediating CD, as has been mentioned at the beginning of this chapter.

Pathological Consequences of CD

In the last decade, the effect of CD on human health has become an important issue. Indeed nowadays we certainly know that several chronic diseases, which widely affect our society, are influenced by chronobiological components (Fig. 4.1).

Chronodisruption and Cancer

Experimentally, it is well documented that unusual changes in the LD cycle that cause chronodisruption lead to accelerated growth of cancers. Using mice bearing transplanted Glasgow osteosarcomas, Filipski et al. [46] noted that these tumors grew more rapidly when the mice were exposed to a simulated eastward flight that caused an 8-h phase advance every 2 days. This recurring unusual 8-h photoperiodic change, which would duplicate a flight from the mid-USA to central Europe caused chronic circadian disruption relative to mice maintained in a stable photoperiodic cycle.

In the study by Filipski and colleagues [46], in addition to the mice suffering circadian misalignment, they very possibly exhibited an abnormal or more likely, no nighttime rise in melatonin production and secretion. Since the actual levels of melatonin were not measured in these animals, it remains undetermined if melatonin suppression was a factor in the more rapid tumor growth. It seems likely, however, that several changes, i.e., circadian disruption, melatonin suppression, sleep deprivation, etc., conspired to promote the accelerated growth of the transplanted osteosarcoma cells.

That electrolytic destruction of the biological clock, i.e., the SCN, and the resultant circadian disruption, stimulates cancer progression was confirmed in a study performed by the same group. In this case, Filipski and Li [47] electrolytically lesioned the SCN of mice bearing implanted tumors. This destructive procedure caused the cancers to grow more rapidly. While loss of the SCN would certainly destroy and/or severely compromise the circadian rhythms of all organs, the lesions also surely demolished the cyclic production and release of pineal melatonin; therefore, the mice were deficient in this cancer-inhibiting indoleamine. Because of the multiple negative actions resulting from lesions of the SCN, the basis of the accelerated progression of the cancer in the mice lacking their central clock remains undetermined.

Other studies also point to a role of the circadian network and chronodisruption in the acceleration of cancer growth [48]. The expression of the *Per2* gene is a critical factor in circadian organization. In mice genetically devoid of the *Per2* gene, the typical circadian cycle of 24–25 h becomes shortened to less than 24 h. A high percentage of the mice with this circadian malady spontaneously develop lymphomas by the time they are 16 months of age. Normally, these tumors in intact mice do not appear until the animals are beyond 20 months of age. The findings of Fu and coworkers [45] also support a role for the involvement of circadian genes in cancer cell proliferation.

Similar to that described in experimental animals, several evidences link CD and cancer also in human. Thus, for at least two decades, epidemiologists have been interested in the association between LAN, along with the associated changes in the underlying physiology, and the elevated incidence of breast cancer risk. The initial reports claimed that a higher frequency of breast cancer was apparent in attendants who routinely worked as airline hostesses on long flights over multiple time zones [49, 50] and in women who commonly performed night shift work over long periods of time [51, 52]. The workers most commonly invoked chronodisruption as a major contributory factor to the reported rise in breast cancer rather than a disturbance or a reduction in the melatonin rhythm [26, 53, 54]. More recently, however, interest in the possible or likely involvement of depressed melatonin levels is attracting greater interest [55, 56]. This surely stems from the experimental studies that document melatonin as a significant endogenous anticancer agent, not only for breast cancer [46, 57, 58] but cancer in general [59–61].

Following the initial observations on breast cancer in females as a consequence of LAN, prostate cancer also began to attract attention. While the studies are not as numerous as those for breast cancer nor is the evidence as provocative, an elevated frequency of prostate cancer may likewise be more frequent in males who experience what would be divergent LD environments [62, 63]. Moreover, it has been speculated that in fact routine perturbations of the LD cycle may be a factor that significantly influences cancer development of many types, not only breast and prostate [59]. Chronodisruption and melatonin suppression as being potentially carcinogenic has even been noted by the World Health Organization, inasmuch as they have classified circadian disturbance as a Group 2A carcinogen; this classification suggests that there is likely (although not definitively proven) a relationship between the elevated frequency of certain cancer types and chronic alterations in the naturally occurring LD cycles [64]. While a considerable amount of data, most of which is observational, suggests that some aspect of circadian disruption contributes to a higher prevalence of certain cancer types, the idea is not enthusiastically supported by some [65]. On the other hand, to assume that routine chronodisruption due to what must be considered abnormal or unusual LD cycles is totally inconsequential in reference to potential pathophysiologies would seem imprudent. Other cancer types that have been tentatively linked to LAN/chronodisruption include endometrial [66], colorectal [67] and lung [68].

If a definitive link between chronic perturbations of the LD cycle and pathologies, i.e., cancer, is established, it will be important to identify the mechanisms involved. With the exposure to light at night, at least three things can happen, i.e., the individual may experience chronodisruption, there may be a reduction in sleep duration or efficiency, and nocturnal melatonin levels may be suppressed (Fig. 4.2). At this point, it is not known which of these contribute to the presumed physiological rearrangements and the alleged pathologies that may occur. Certainly each of these, i.e., chronodisruption, sleep deprivation, and melatonin suppression, may individually be capable of contributing to any pathology that occurs. It would seem what is most likely, however, is that these unusual disturbances conspire to precipitate molecular physiological perturbations at the level of individual cells that then eventually cause overt pathologies.

Interestingly, the changes caused by light exposure at abnormal times, i.e., during darkness, cause functional changes that are reminiscent of those seen in older humans, when pathologies of various types are more likely to be manifested. Thus, aging is associated with changes typical of chronodisruption [1], sleep is often impaired in the elderly [15, 21, 38, 39], and melatonin levels diminish with advanced age [69, 70].

Finally, if cancer, particularly breast cancer, is one manifestation of excessive and chronic exposure to light at night, it would seem likely that this would not be the only pathology that would occur. Indeed, within the last decade many diseases have been at least theoretically related to disturbances of biological rhythmicity, sleep deprivation and less than adequate amounts of sleep [29, 38, 71–73]. Thus, it would seem likely that many pathologies which are common in the aged, may in fact also be aggravated by chronic exposure to excessive light at night (Fig. 4.2).

Fig. 4.2 (continued) period (night shift work or light pollution/trespass light). Light of adequate intensity and wavelength is detected by the intrinsic photoreceptive retinal ganglion cells (*ip*RGCs), melanopsin containing neurons, with the resulting signal being sent via the axons (the retinohypothalamic tract) of these cells to the biological clock, i.e., the suprachiasmatic nuclei (SCN). Clock disturbances can then be transferred, via neural or humoral signals, to all cells in the body, all of which possess genes that are under circadian regulation. The resulting disturbances in cell physiology may culminate in functional disturbances that lead to a variety of pathologies. One challenge for scientists/clinicians is to definitively establish whether there exist serious pathologies that relate to light at night and, if so, to clarify the mechanisms involved so corrective actions can be instituted

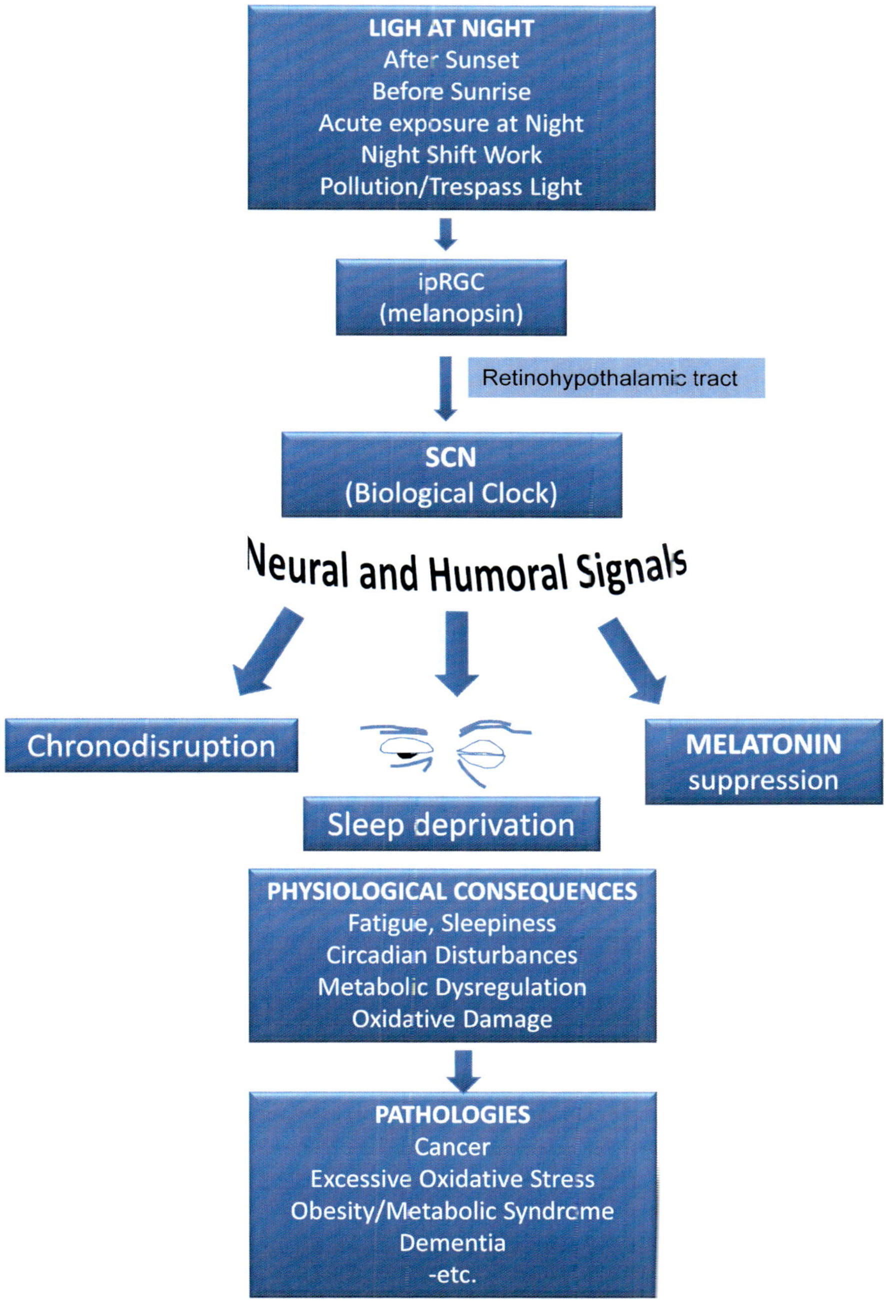

Fig. 4.2 A flow diagram illustrating the sequence of events that may lead to pathologies reportedly associated with light at night. Light at night may take many forms such as reducing the duration of the dark period (light after sunset and before sunrise), transitory interruption of the period of darkness (acute light exposure at night), and total or near total elimination of the daily dark

Chronodisruption, Metabolic Syndrome (MetS), and Obesity

Since the discovery of the CLOCK mutant mice, to the present moment many outstanding and consistent studies have demonstrated the important connection between CD and obesity (Fig. 4.3).

Although the connection between obesity and CD will be treated in several chapters along the present book, it is important to highlight that our experiments in humans are showing that MetS metabolic disturbances such as increased blood pressure, increased glucose and plasma lipids regulation, and changes in adipocyte-secreted hormones such as leptin and adiponectin, associates with diminished daily amplitude in melatonin and cortisol circadian patterns, demonstrated the existence of chronodisruption with metabolic syndrome [74]. In this line, in another experiment also performed by our group, analyses of skin temperature indicated that obese women displayed significantly lower mean wrist temperature (WT) with a more flattened 24-h pattern, a lower-quality rhythm, and a higher intradaily variability. Particularly interesting were the marked differences between obese and normal-weight women in the secondary WT peak in the postprandial period, considered as a marker of chronodisruption and of metabolic alterations. These 24-h changes were associated with higher MetS risk [75].

Other Pathologies Related to CD

While this chapter considered mainly the potential association of chronodisruption with cancer and with obesity, there may well be other health-related problems that occur as a consequence of imposed chronic changes in the LD environment that severely disturb circadian rhythmicity [76].

Multiple records link CD with the increased risk of developing premature aging, cardiovascular diseases, cognitive impairment and mood disorders, among others.

Similar to that observed in many physiological processes, the functioning of the circadian system changes with *age*. Phase advance, reduced amplitude, circadian fragmentation, impaired ability to resynchronize after a time shift and internal desynchronization among different rhythms are the major characteristics of aged rhythms [77]. However, less well known is the fact that CD has a direct role in inducing accelerated aging [78, 79]. Although it is generally believed that disruptions in circadian rhythms lead to reduced life expectancy, whereas their appropriate resetting leads to well-being and increased longevity [77].

Cardiovascular diseases are also related to chronodisruption. In people with normal blood pressure (BP) and uncomplicated essential hypertension, BP declines to its lowest levels during night-time sleep, rises abruptly with morning awakening and attains a maximum during diurnal activity. It has been shown that night-time BP is the best predictor of stroke and myocardial infarction risk [80, 81]. Thus, hypertensive patients with a normal reduction in nocturnal BP (dipper) had a relative

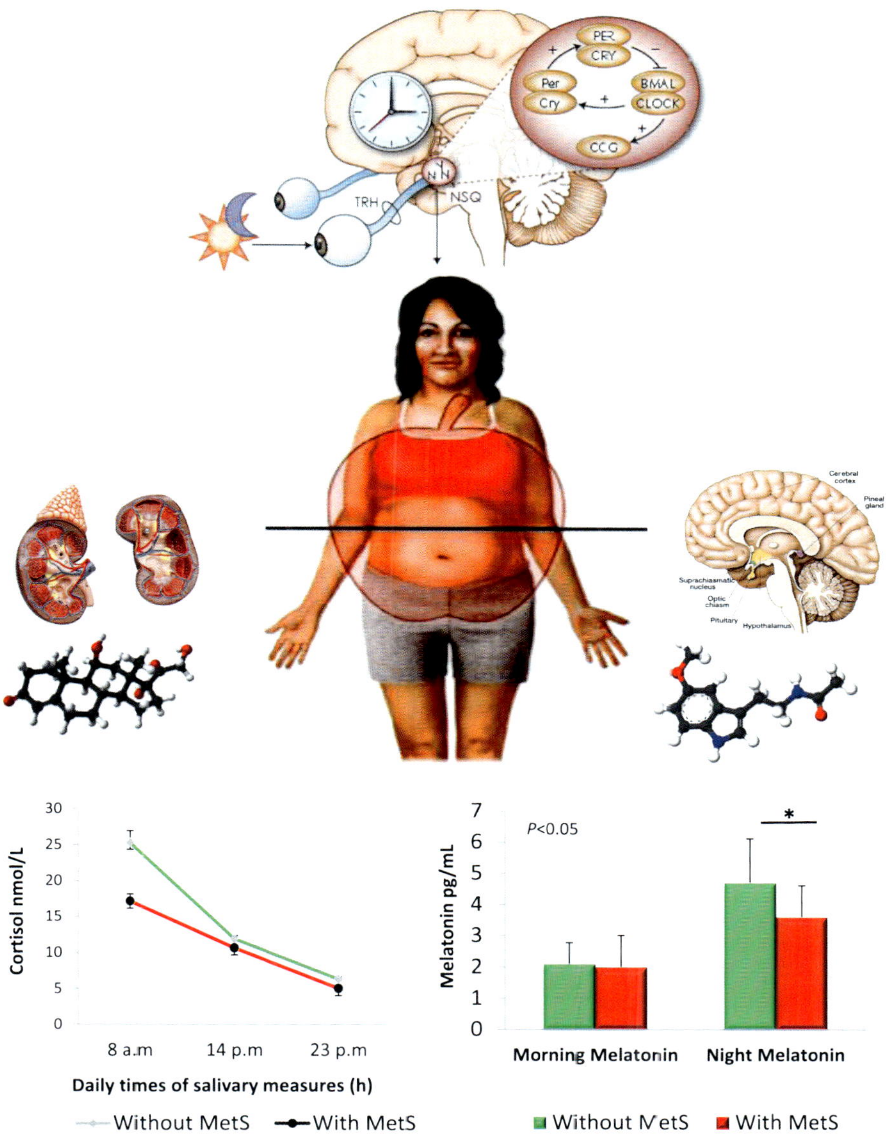

Fig. 4.3 Relationship between circadian disruption and obesity. The photic inputs that reach the SCN are processed and forwarded to a small number of other hypothalamic nuclei and to the adrenal and pineal gland to regulate cortisol and melatonin rhythms. Melatonin is the end product of a biosynthetic pathway that begins with the nutrient amino acid tryptophan The relationship between light and melatonin is inverse. When the SCN is stimulated by daylight signals from the retina, it instructs the pineal gland to suppress melatonin production. Then, when daylight fades in the evening, melatonin secretion is increased many times over, creating a physiological condition of "biological darkness" in the person. The circadian rhythm of cortisol and melatonin secretion together with sleep–wake rhythm are impaired in obese people. Morning cortisol levels and evening melatonin levels are reduced in obese compared with control women (Data from Corbalán-Tutau et al. 2011. Chronobiol. Int. 28:425–433)

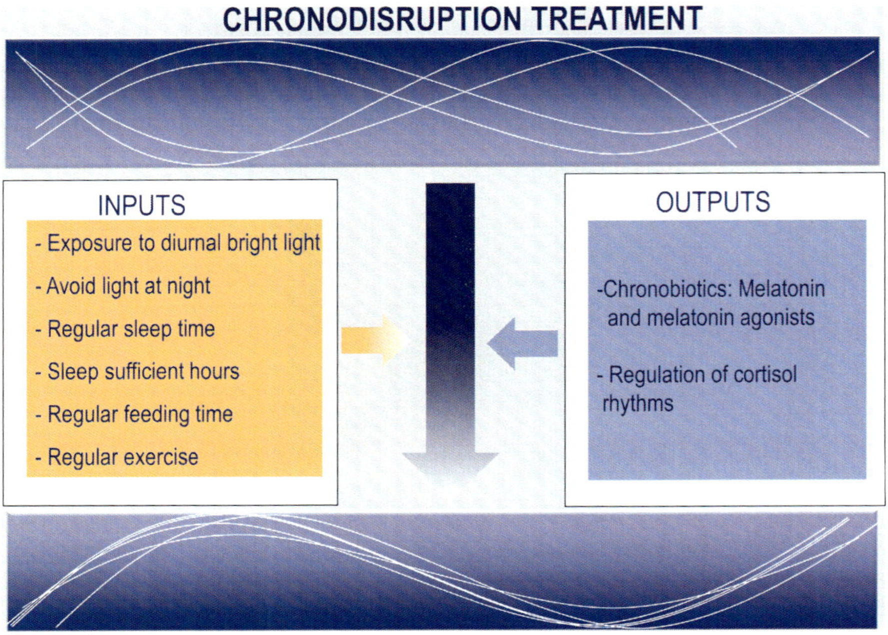

Fig. 4.4 Strategies for circadian enhancement. The adequate functioning of the circadian system is dependent on regularly timed exposure to synchronizers. Regarding the outputs, regularly timed melatonin treatment enhances circadian robustness as occurred with a stable cortisol rhythm

hazard of cardiovascular mortality similar to that in non-dipper normotensives. It is noteworthy that the non-dipper circadian pattern is more frequent among shift workers and elderly people [80, 81].

The circadian system modulates cognition and affective function by projections from SCN to arousal and sleep systems and through eliciting changes in clock genes in extra-SCN brain regions. Chronodisruption in the sleep–wake circadian rhythm has important implications for learning, memory and emotion [82]. When wakefulness occurs at appropriate internal biological times, circadian system benefits human cognitive and emotion function. However, when wakefulness occurs at inappropriate biological times because of social pressures, such as early school start times, work at night, shift work, and jet lag, or because of circadian sleep disorders, the resulting misalignment between circadian and wakefulness–sleep physiology leads to impaired cognitive performance, learning, emotion, and safety [82].

Experimental studies specifically addressing the treatment of chronodisruption are lacking. However, the development of technical improvement in healthy lighting systems and all behavioral and pharmacological treatments that improve the circadian system status (chronoenhancement) may help to reduce the risk of pathologies associated to chronodisruption (Fig. 4.4).

Concluding Remarks

As discussed in this brief review, the misuse of artificial light, i.e., light after darkness onset, causes disintegration of biological clock processes such that the temporal architecture of cells, which is the basis of optimal physiology, breaks down, thereby contributing to pathophysiologies. Additionally, LAN reduces an important chemical messenger from the pineal gland, melatonin, the loss of which likely also contributes to increased pathologies, e.g., cancer, and reduced quality of life. While this brief review considered the potential association of chronodisruption and cancer, there may well be other health-related problems that occur as a consequence of imposed chronic changes in the LD cycle and other non photic synchronizers that severely disturb circadian rhythmicity [76].

Summary Points

- The suprachiasmatic nucleus conveys circadian information to the peripheral cellular oscillators through both neural and humoral messages. One important humoral message is the melatonin signal from the pineal gland.
- Melatonin, also known as "chemical darkness," shows a circadian rhythm which is regulated by a double mechanism: an endogenous pattern driven by the suprachiasmatic nucleus of hypothalamus and an acute inhibitory effect of nocturnal light.
- The introduction of artificial sources of light roughly 150 years ago has significantly undermined the naturally occurring light–dark cycle and, likewise, has disturbed circadian rhythms since light is now available at unusual times. Light at night is known to cause circadian disruption and melatonin suppression.
- In our modern society, circadian disruption may be common in several conditions such as jet lag, shift work, light at night, or social jet lag. In addition clock gene polymorphisms and aging may also predispose to chronodisruption.
- In the recent years chronodisruption is being associated with an impairment of several pathologies such as cancer, cardiovascular diseases, sleep disorders, premature aging, obesity, metabolic syndrome, and cognitive and affective disorders.

References

1. Erren TC, Reiter RJ (2009) Defining chronodisruption. J Pineal Res 46:245–247
2. Garaulet M, Ordovas JM, Gomez-Abellan F, Martinez JA, Madrid JA (2011) An approximation of the temporal order in endogenous circadian rhythms of genes implicated in human adipose tissue metabolism. J Cell Physiol 226:2075–2080

3. Hardeland R, Madrid JA, Tan DX, Reiter RJ (2012) Melatonin, the circadian oscillatory system and health: the need for detailed analyses of peripheral melatonin signaling. J Pineal Res 52:139–166
4. Do MT, Yau KW (2010) Intrinsically photosensitive retinal ganglion cells. Physiol Rev 90:1547–1581
5. Chen SK, Badea TC, Hattar S (2011) Photoentrainment and pupillary light reflex are mediated by distinct populations of ipRGCs. Nature 476:92–95
6. Sand A, Schmidt TM, Kofuji P (2012) Diverse types of ganglion cell photoreceptors in the mammalian retina. Prog Retin Eye Res 31(4):287–302
7. Stehle JH, Saade A, Rawashdeh O, Akermann K, Jilg A, Sebesteny T, Maronde E (2011) A survey of molecular details in the human pineal gland in the light of phylogeny, structure, function and chronobiological diseases. J Pineal Res 51:17–43
8. Panke ES, Rollag MD, Reiter RJ (1980) Effects of photoperiod on hamster pineal melatonin concentrations. Comp Biochem Physiol 66A:691–693
9. Vaughan GM, Pelham RW, Pang SF, Laughlin LL, Wilson KW, Sandock KL, Vaughan MK, Koslaw SH, Reiter RJ (1976) Nocturnal elevation of plasma melatonin and urinary 5-hydroxyindoleacetic acid in young men: attempts at modification by brief changes in environmental lighting and sleep and by autonomic drugs. J Clin Endocrinol Metab 42:752–764
10. Jung-Hynes B, Reiter RJ, Ahmad N (2010) Sirtuins, melatonin and circadian rhythms: building a bridge between aging and cancer. J Pineal Res 48:9–19
11. Garaulet M, Ordovás JM, Madrid JA (2010) The chronobiology, etiology and pathophysiology of obesity. Int J Obes (Lond) 34(12):1667–1683, Epub 2010 Jun 22. Review
12. Reiter RJ, Paredes SD, Manchester LC, Tan DX (2009) Reducing oxidative/nitrosative stress: a newly-discovered genre for melatonin. Crit Rev Biochem Mol Biol 44:175–200
13. Martinez-Nicolas A, Ortiz-Tudela E, Madrid JA, Rol MA (2011) Cross talk between environmental light and internal time in humans. Chronobiol Int 28:617–629
14. Lewy AJ, Wehr TA, Goodwin FK, Newsome DA, Markey SP (1980) Light suppresses melatonin secretion in humans. Science 201:1267–1269
15. Brainard GC, Richardson BA, Petterbarg LJ, Reiter RJ (1982) The effect of different light intensities on pineal melatonin content. Brain Res 233:75–81
16. Brainard GC, Richardson BA, King TS, Matthews SA, Reiter RJ (1983) The suppression of pineal melatonin content and N-acetyltransferace activity by different light irradiances: a dose–response relationship. Endocrinology 113:293–296
17. Brainard GC, Richardson BA, Hurlbut EC, Steinlechner S, Matthews SA, Reiter RJ (1984) The influence of various irradiances of artificial light, twilight and moonlight on the suppression of pineal melatonin content in the Syrian hamster. J Pineal Res 1:105–119
18. Brainard GC, Richardson BA, King TS, Reiter RJ (1984) The influence of different light spectra in the suppression of pineal melatonin content in the Syrian hamster. Brain Res 294:333–339
19. Wehr TA (1991) The duration of human melatonin secretion and sleep respond to changes in daylength (photoperiod). J Clin Endocrinol Metab 73:1276–1280
20. Bass J, Turek FW (2005) Sleepless in America: a pathway to obesity and metabolic syndrome. Arch Intern Med 165:15–16
21. Reiter RJ, Tan DX, Erren TC, Fuentes-Broto L, Paredes SD (2009) Light-mediated perturbations of circadian timing and cancer risk: a mechanistic analysis. Integr Cancer Ther 8:354–360
22. Tan DX, Chen LD, Poeggeler B, Manchester LC, Reiter RJ (1993) Melatonin: a potent, endogenous hydroxyl radical scavenger. Endocr J 1:57–60
23. Blask DE, Brainard GC, Dauchy RT, Hanifin JP, Davidson LK, Krause JA, Sauer LA, Rivera-Bermudez MA, Dubocovich ML, Jasser SA, Lynch DT, Rollag MD, Zalatan F (2005) Melatonin-depleted blood from premenopausal women exposed to light at night stimulates growth of human breast cancer xenografts in nude rats. Cancer Res 65:11174–11184
24. Reiter RJ, Tan DX, Fuentes-Broto L (2010) Melatonin: a multi-tasking molecule. Prog Brain Res 181:127–151

25. Galano A, Tan DX, Reiter RJ (2011) Melatonin as a natural ally against oxidative stress: a physiochemical examination. J Pineal Res 51:1–16
26. Hanson J (2001) Increased breast cancer risk among women who work predominantly at night. Epidemiology 12:74–77
27. Erren TC, Pepe HG, Reiter RJ, Prekarski C (2008) Chronodisruption and cancer. Naturwissenschaften 95:367–382
28. Alvarez-Garcia V, Gonzalez A, Alonso-Gonzalez C, Martinez-Campa C, Cos S (2012) Melatonin interferes in the desmoplastic reaction in breast cancer by regulating cytokine production. J Pineal Res 52:292–290
29. Turek FW, Joshu C, Kahsaka A, Lin E, Ivanova G, McDeamon E, Laposky A, Losee-Olson S, Easton A, Jensen DR, Eckel RH, Takahashi JS, Bass J (2005) Obesity and metabolic syndrome in circadian Clock mutant mice. Science 308:1043–1045
30. Reiter RJ, Tan DX, Korkmaz A, Erren TC, Piekarski C, Tamura H, Manchester LC (2007) Light-at-night, chronodisruption, melatonin suppression and cancer risk: a review. Crit Rev Oncog 13:303–328
31. Mediavilla MD, Sanchez-Barcelo E, Tan DX, Manchester LC, Reiter RJ (2010) Basic mechanism involved in the anti-cancer actions of melatonin. Curr Med Chem 17:4462–4481
32. Blask DE, Hill SM, Dauchy RT, Xiang S, Yuan L, Duplessis T, Mao L, Dauchy E, Sauer LA (2011) Circadian regulation of molecular, dietary, and metabolic signaling mechanisms of human breast cancer growth by the nocturnal melatonin signal and the consequences of its disruption by light at night. J Pineal Res 51:259–269
33. Van Cauter E, Spiegel K, Tasali E, Leproult R (2008) Metabolic consequences of sleep and sleep loss. Sleep Med 9(Suppl 1):S23–S28
34. National Center for Health Statistics (2000) Quick stats: percentage of adults who report an average of 6 hours of sleep per 24-hour period by sex and age group United States, 1985–2004. Biol Psychiatr 47:921–927
35. Verkasalo PK, Lillberg K, Stevens RG, Hublin C, Partinen M, Koskenvuo M, Kaprio J (2005) Sleep duration and breast cancer: a prospective cohort study. Cancer Res 65:9595–9606
36. Pinheiro SP, Schernhammer ES, Tworoger SS, Michels KB (2006) A prospective study of habitual duration of sleep and incidence of breast cancer in a large cohort of women. Cancer Res 66:5521–5525
37. Wu AH, Wang R, Koh WP, Stanczyk FZ, Lee HP, Wu MC (2008) Sleep duration, melatonin and breast cancer among Chinese women in Singapore. Carcinogenesis 29:1244–1248
38. Blask DE (2009) Melatonin, sleep disturbances and cancer risk. Sleep Med Rev 13:357–364
39. Hill SM, Frasch T, Xiang S, Yuan I, Duplessis T, Mao L (2009) Molecular mechanisms of melatonin anticancer effects. Integr Cancer Ther 8:301–308
40. Stokkan KA, Yamazaki S, Tei H, Sakaki Y, Menaker M (2001) Entrainment of the circadian clock in the liver by feeding. Science 291:490–493
41. Qin LQ, Li J, Wang Y, Wang J, Xu JY, Kaneko T (2003) The effects of nocturnal life on endocrine circadian patterns in healthy adults. Life Sci 73:2467–2475
42. Scheer FA, Hilton MF, Mantzoros CS, Shea SA (2009) Adverse metabolic and cardiovascular consequences of circadian misalignment. Proc Natl Acad Sci U S A 106:4453–4458
43. Bechtold DA, Gibbs JE, Loudon AS (2010) Circadian dysfunction in disease. Trends Pharmacol Sci 31:191–198
44. Kondratov RV, Kondratova AA, Gorbacheva VY, Vykhovanets OV, Antoch MP (2006) Early aging and age-related pathologies in mice deficient in BMAL1, the core component of the circadian clock. Genes Dev 20:1868–1873
45. Fu L, Pelicano H, Liu J, Huang P, Lee C (2002) The circadian gene Period 2 plays an important role in tumor suppression and DNA damage response in vivo. Cell 111:41–50
46. Filipski E, Dalauney F, King VM (2004) Effects of chronic jet lag in tumor progression in mice. Cancer Res 64:7879–7985
47. Filipski E, Li XM (2006) Disruption of circadian coordination and tumor progression in mice. Cancer Causes Control 17:509–514
48. Rosbach M, Takahachi JS (2002) The cancer connection. Nature 420:373–374

49. Rafnsson V, Tulinius H, Jonasson JG, Hrafnkelsson J (2001) Risk of breast cancer in female flight attendants: a population-based study (Iceland). Cancer Causes Control 12:95–101
50. Kojo K, Pukkala F, Auvinen A (2005) Breast cancer risk among Finnish cabin attendants: a nested case–control study. Occup Environ Med 62:488–493
51. Megdal SP, Kroenke CN, Laden F, Pukkala E, Schernhammer ES (2005) Night shift work and breast cancer: a systematic review and meta-analysis. Eur J Cancer 41:2023–2032
52. Stevens RG (2005) Circadian disruption and breast cancer: from melatonin to clock genes. Epidemiology 16:254–258
53. Davis S, Mirick DK, Stevens RG (2001) Night shift work, light at night and risk of breast cancer. J Natl Cancer Inst 93:1557–1562
54. Kolstad HA (2008) Shift work and the risk of breast cancer and other cancers – a critical review of the epidemiologic evidence. Scand J Work Environ Health 34:5–22
55. Viswanathan AN, Schernhammer ES (2009) Circulating melatonin and the risk of breast and endometrial cancer in women. Cancer Lett 28:1–7
56. Davis S, Mirick DK, Chen C, Stanczyk FZ (2012) Night shift work and hormone levels in women. Cancer Epidemiol Biomarkers Prev 21:609–618
57. Lee SE, Kim SJ, Youn JP, Hwang SY, Park CS, Park YS (2011) MicroRNA and gene expression analysis of melatonin-exposed human breast cancer cell lines indicating involvement of the anticancer effect. J Pineal Res 51:345–352
58. Proietti S, Cucina A, D'Anselmi F, Dinicola S, Pasqualato A, Lisi E, Bizzari M (2011) Melatonin and vitamin D3 synergistically down-regulate Akt and MDM2 leading to TGFβ-1-dependent growth inhibition of breast cancer cells. J Pineal Res 50:150–158
59. Erren TC, Reiter RJ (2008) A generalized theory of carcinogenesis due to chronodisruption. Neuro Endocrinol Lett 29:815–821
60. Carbajo-Pescador S, Garcia-Palomo A, Martin-Renedo J, Piva M, Gonzalez-Gallego J, Mauriz JL (2011) Melatonin modulation of intracellular-signaling pathways in hepatocarcinoma HepG2 cell line: role of the MT1 receptor. J Pineal Res 51:463–471
61. Cho SY, Lee HJ, Jeong SJ, Lee HJ, Kim HS, Chen CY, Lee EO, Kim SH (2001) Sphingosine kinase 1 pathway is involved in melatonin-induced HIF-1α inactivation in hypoxic PC-3 prostate cancer cells. J Pineal Res 51:87–93
62. Kubo T, Ozasa K, Mikami K, Wakai K, Fujino Y, Watanabe Y, Miki T, Nakao M, Hayaski K, Suzuki K, Mori M, Washio M, Sakauchi F, Ito Y, Yoshimura T, Tamakoshi A (2006) Prospective cohort study of the risk of prostate cancer among rotating shift workers: findings from the Japan collaborative cohort study. Am J Epidemiol 164:549–555
63. Costa G, Haus E, Stevens R (2010) Shift work and cancer—consideration on rationale, mechanisms, and epidemiology. Scand J Work Environ Health 36:163–179
64. Straif K, Baan R, Grosse Y, Secretan B, El-Ghissassi F, Bouvard V, Altieri A, Benbrahim-Tallaa L, Cogliano V (2007) Carcinogenicity of shift work, painting and fire-fighting. Lancet Oncol 8:1065–1066
65. Kantermann T, Roenneberg T (2009) Is light-at-night a health risk factor or health risk predictor? Chronobiol Int 6:1069–1074
66. Sturgeon SR, Luisi N, Balasubramanian R, Reeves KW (2012) Sleep duration and endometrial cancer risk. Cancer Causes Control 23:547–553
67. Schernhammer ES, Laden F, Speizer FE, Willett WC, Hunter DJ, Kawachi I, Fuchs CS, Colditz GA (2003) Night-shift work and risk of colorectal cancer in the nurse' health study. J Natl Cancer Inst 95:825–828
68. Logan RW, Zhang C, Murugan S, O'Connell S, Levitt D, Rosenwasser AM, Sarkar DK (2012) Chronic shift-lag alters the circadian clock of NK cells and promotes lung cancer growth in rats. J Immunol 188:2583–2591
69. Sack RL, Lewy AJ, Erb DL, Vollmer WM, Singer CM (1986) Human melatonin production decreases with age. J Pineal Res 3:379–388
70. Reiter RJ, Richardson BA, Johnson LY, Ferguson BN, Dinh DT (1980) Pineal melatonin rhythm: reduction in aging Syrian hamsters. Science 210:1372–1373

71. Tan DX, Manchester LC, Fuentes-Broto L, Paredes SD, Reiter RJ (2011) Significance and application of melatonin in the regulation of brown adipose tissue metabolism: relation to human obesity. Obes Rev 12:167–188
72. Tan CW, Shiu SYW (2011) Functional interplay between melatonin-receptor-mediated antiproliferative signaling and androgen receptor signaling in human prostate epithelial cells: potential implications for therapeutic strategies against prostate cancer. J Pineal Res 51:297–312
73. Reiter RJ, Tan DX, Korkmaz A, Ma S (2012) Obesity and metabolic syndrome: association with chronodisruption, sleep deprivation, and melatonin suppression. Ann Med 44:564–577
74. Corbalan-Tutau MD, Madrid JA, Nicolas F, Garaulet M (2012) Daily profile in two circadian markers "melatonin and cortisol" and associations with Metabolic Syndrome components. Physiol Behav. Jun 13 [Epub ahead of print]
75. Corbalán-Tutau MD, Madrid JA, Ordovás JM, Smith CE, Nicolás F, Garaulet M (2011) Differences in daily rhythms of wrist temperature between obese and normal-weight women: associations with metabolic syndrome features. Chronobiol Int 28:425–433
76. Flo E, Pallesen S, Mageroy N, Moen BE, Gronli J, Nordhus IH, Bjorvatn B (2012) Shift work disorders in nurses—assessment, prevalence and related health problems. PLoS One 7:e33981
77. Hofman MA, Swaab DF (2006) Living by the clock: the circadian pacemaker in older people. Ageing Res Rev 5:33–51
78. Davidson AJ, Sellix MT, Daniel J, Yamazaki S, Menaker M, Block GD (2006) Chronic jet-lag increases mortality in aged mice. Curr Biol 16:R914–R916
79. Hurd MW, Ralph MR (1998) The significance of circadian organization for longevity in the golden hamster. J Biol Rhythms 13:430–436
80. Staessen JA, Fagard R, Thijs L, Celis H, Arabidze GG, Birkenhäger WH, Bulpitt CJ, de Leeuw PW, Dollery CT, Fletcher AE, Forette F, Leonetti G, Nachev C, O'Brien ET, Rosenfeld J, Rodicio JL, Tuomilehto J, Zanchetti A (1997) Randomised double-blind comparison of placebo and active treatment for older patients with isolated systolic hypertension. The Systolic Hypertension in Europe (Syst-Eur) Trial Investigators. Lancet 350:757–764
81. Ohkubo T, Hozawa A, Yamaguchi J, Kikuya M, Ohmori K, Michimata M, Matsubara M, Hashimoto J, Hoshi H, Araki T, Tsuji I, Satoh H, Hisamichi S, Imai Y (2002) Prognostic significance of the nocturnal decline in blood pressure in individuals with and without high 24-h blood pressure: the Ohasama study. J Hypertens 20:2183–2189
82. Wright KP, Lowry CA, Lebourgeois MK (2012) Circadian and wakefulness-sleep modulation of cognition in humans. Front Mol Neurosci 5:50, Epub 2012 Apr 18

Chapter 5
Obesity and Chronodisruption: An Imbalance Between Energy Intake and Expenditure

Oren Froy

Abstract Obesity has become a serious public health problem and a major risk factor for the development of illnesses, such as insulin resistance and hypertension. Attempts to understand the causes of obesity and develop new therapeutic strategies have mostly focused on caloric intake and energy expenditure. Recent studies have shown that the circadian clock controls energy homeostasis by regulating circadian expression and/or activity of enzymes, hormones, and transport systems involved in metabolism. Moreover, disruption of circadian rhythms leads to obesity and metabolic disorders. Therefore, it is plausible that resetting of the circadian clock can be used as a new approach to attenuate obesity. Feeding regimens, such as restricted feeding (RF), calorie restriction (CR) and intermittent fasting (IF), provide a time cue and reset the circadian clock and lead to better health. In contrast, high-fat (HF) diet leads to disrupted circadian expression of metabolic factors and obesity. This chapter will focus on chronodisruption and feeding regimens with implications for obesity.

Introduction: Obesity and Chronodisruption

Obesity has become a serious and growing public health problem [1]. Attempts to understand the causes of obesity and develop new therapeutic strategies have mostly focused on the imbalance between energy expenditure and caloric intake. Recent studies link energy regulation to the circadian clock at the behavioral, physiological, and molecular levels [2–5], emphasizing that the timing of food intake itself may

O. Froy, Ph.D. (✉)
The Robert H. Smith Faculty of Agriculture, Food and Environment, Nutrigenomics and Functional Foods Research Center, Institute of Biochemistry, Food Science and Nutrition, The Hebrew University of Jerusalem, P.O. Box 12, Rehovot 76100, Israel
e-mail: oren.froy@mail.huji.ac.il

M. Garaulet and J.M. Ordovás (eds.), *Chronobiology and Obesity*,
DOI 10.1007/978-1-4614-5082-5_5,

play a significant role in weight gain [6]. The mammalian circadian clock influences nearly all aspects of physiology and behavior, including sleep–wake cycles, cardiovascular activity, endocrine system, body temperature, renal activity, physiology of the gastrointestinal tract, and hepatic metabolism [7]. Disruption of circadian coordination (chronodisruption) may be manifested by hormone imbalance, psychological and sleep disorders, cancer proneness, and reduced life span [7–11]. In contrast, robust circadian rhythms lead to well-being and increased longevity [12, 13]. This correlation reveals the prominent influence of the circadian clock on human physiology and pathophysiology. This chapter will summarize recent findings concerning the relationship between circadian rhythms, food intake, and energy expenditure with implications for obesity.

Circadian Rhythms in Metabolism

The circadian clock regulates metabolism and energy homeostasis in peripheral tissues [5, 14]. This is achieved by mediating the expression and/or activity of certain metabolic enzymes and transport systems [15, 16] involved in cholesterol metabolism, amino acid regulation, drug and toxin metabolism, the citric acid cycle, and glycogen and glucose metabolism [5, 14, 17–20]. Moreover, lesion of rat central clock in the brain suprachiasmatic nuclei (SCN) abolishes daily variations in whole body glucose homeostasis [21], altering not only rhythms in glucose utilization rates but also endogenous hepatic glucose production. Indeed, the SCN projects to the pre-autonomic paraventricular nucleus (PVN) neurons to control hepatic glucose production [22]. Similarly, glucose uptake and the concentration of the primary cellular metabolic currency adenosine triphosphate (ATP) in the brain and peripheral tissues have been found to fluctuate around the circadian cycle [18, 22, 23].

Many hormones involved in metabolism, such as insulin [17], glucagon [24], adiponectin [25], corticosterone [26], leptin, and ghrelin [27, 28], have been shown to exhibit circadian oscillation. Leptin, an adipocyte-derived circulating hormone that acts at specific receptors in the hypothalamus to suppress appetite and increase metabolism, is extremely important in obesity. Plasma leptin levels are normally pulsatile and circadian with leptin peaking early in the non-active phase, that is during the early dark phase in diurnal animals, such as monkeys and humans [29, 30], and during the early to mid-light phase in nocturnal animals, such as rats and mice [31, 32]. Neither feeding time nor adrenalectomy affected the rhythmicity of leptin release. However, ablation of the suprachiasmatic nuclei (SCN), the location of the circadian clock in the hypothalamus, was shown to eliminate leptin circadian rhythmicity in rodents, suggesting that the central circadian clock regulates leptin expression [31]. In addition, SCN-lesioned rats, as opposed to intact animals, showed no elevation in plasma free fatty acids after intraperitoneal administration of leptin, suggesting a role for SCN in leptin function [33]. In obese subjects, leptin retains diurnal variation in release, but with lower amplitude [34]. Leptin 24-h levels were lower in obese compared with non-obese adolescent girls, suggesting that blunted circadian variation may play a role in leptin resistance and obesity [35].

Receptors for leptin are present on SCN cells [36–38], so it is possible that leptin binds directly to SCN neurons. Thus, it seems that leptin may affect the SCN directly and/or through its effect on the arcuate nucleus (ARC), an aggregation of neurons in the mediobasal hypothalamus important in the regulation of appetite, which is then relayed to the SCN. This regulation of leptin expression by the circadian clock as well as the possible effect of leptin on the SCN places leptin as a possible bridge between energy homeostasis and circadian control. Homozygous C57BL/6 J *Clock*$^{\Delta 19}$ mice, with a truncated exon 18 and deleted exon 19 of the *Clock* gene, that have a greatly attenuated diurnal feeding rhythm, are hyperphagic and obese, and develop a metabolic syndrome of hyperleptinemia, hyperlipidemia, hepatic steatosis, and hyperglycemia [3]. Combination of the *Clock*$^{\Delta 19}$ mutation with the leptin knockout (*ob/ob*) resulted in significantly heavier mice than the *ob/ob* phenotype [39], reiterating the interrelations between leptin and the circadian clock [5, 14, 40].

Circadian Rhythms and Body Weight

Fluctuations in body weight have been associated with changes in day length in various species, suggesting a central role for the circadian clock in regulating body weight. For example, in Siberian hamsters, modulation of body weight depends on photoperiod acting via the temporal pattern of melatonin secretion from the pineal gland, allowing the entrainment of the circadian rhythms of several biological functions [41, 42]. In studies performed on sheep, adipose tissue leptin levels were modulated by day length independently of food intake, body fatness, and gonadal activity. In addition, increasing the length of the photoperiod resulted in increased activity of the lipogenesis-promoting proteins lipoprotein lipase and malic enzyme, independent of the nutritional status [43, 44]. In humans, studies have demonstrated an increased incidence of obesity among shift workers [45–47].

In obese subjects, leptin retains diurnal variation in release, but with lower amplitude, the magnitude of change in the oscillation [34]. Leptin 24-h levels were lower in obese compared with non-obese adolescent girls, suggesting that blunted circadian variation may play a role in leptin resistance and obesity [35]. Circadian patterns of leptin concentration were distinctly different between adult women with upper-body or lower-body obesity, with a delay in peak values of leptin of approximately 3 h in women with upper-body obesity [48]. Indeed, leptin and the leptin receptor knockouts in animals or mutations in humans have been demonstrated to produce morbid, early onset obesity, hypoleptinemia, hyperphagia, hyperinsulinemia, and hyperglycemia [49, 50]. Similarly to leptin, the rhythmic expression of resistin and adiponectin, two other cytokines secreted from adipose tissue, was greatly blunted in obese (KK) and obese, diabetic (KK-A^{y}) mice [25]. In humans, circulating adiponectin levels exhibit both ultradian pulsatility and a diurnal variation. In the latter case, the pattern of adiponectin release is out of phase with leptin

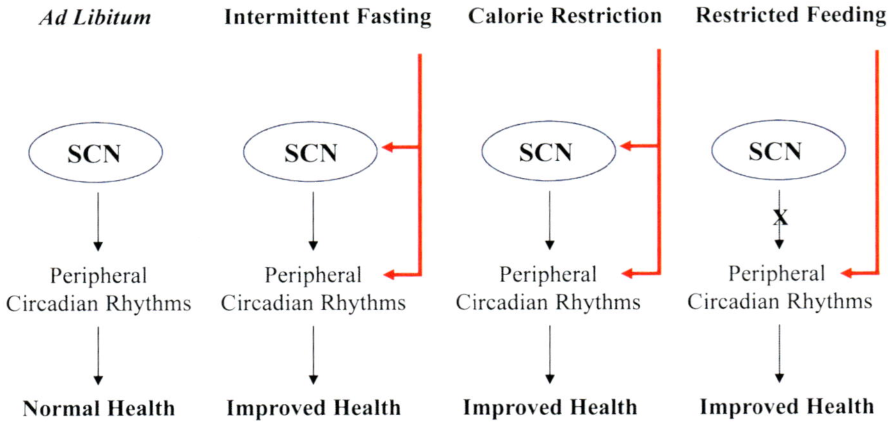

Fig. 5.1 Effect of feeding regimens on circadian rhythms and health. *SCN* suprachiasmatic nuclei

with a significant decline at night, reaching a nadir in the early morning [51]. In obese subjects, adiponectin levels were significantly lower than lean controls, although the obese group had significantly higher average peak of secretion [52]. In rats, melatonin, a synchronizer of the SCN clock, decreased weight gain in response to high-fat diet and decreased plasma leptin levels within 3 weeks. These effects were independent of total food consumption [53]. Thus, it seems that the circadian clock plays a major role in determining body weight probably by influencing the expression and secretion of hormones. Similarly to the control of the circadian clock on metabolism, feeding is a very potent synchronizer (*zeitgeber*) for peripheral clocks (Fig. 5.1).

Effect of Restricted Feeding (RF) on Circadian Rhythms

Limiting the time and duration of food availability with no calorie reduction is termed restricted feeding (RF) [15, 54–56]. Animals, which receive food ad libitum everyday at the same time for only a few hours, adjust to the feeding period within a few days and consume their daily food intake during that limited time [57–59]. Restricting food to a particular time of day has profound effects on the behavior and physiology of animals. 2–4 h before the meal, the animals display food anticipatory behavior, which is demonstrated by an increase in locomotor activity, body temperature, corticosterone secretion, gastrointestinal motility, and activity of digestive enzymes [54, 57, 60, 61], all are known output systems of the biological clock. RF is dominant over the SCN and drives rhythms in arrhythmic and clock mutant mice and animals with lesioned SCN, regardless of the lighting conditions [54, 62–66]. In most incidents, RF affects circadian oscillators in peripheral tissues, such as liver, kidney, heart, and pancreas, with no effect on the central pacemaker in the SCN

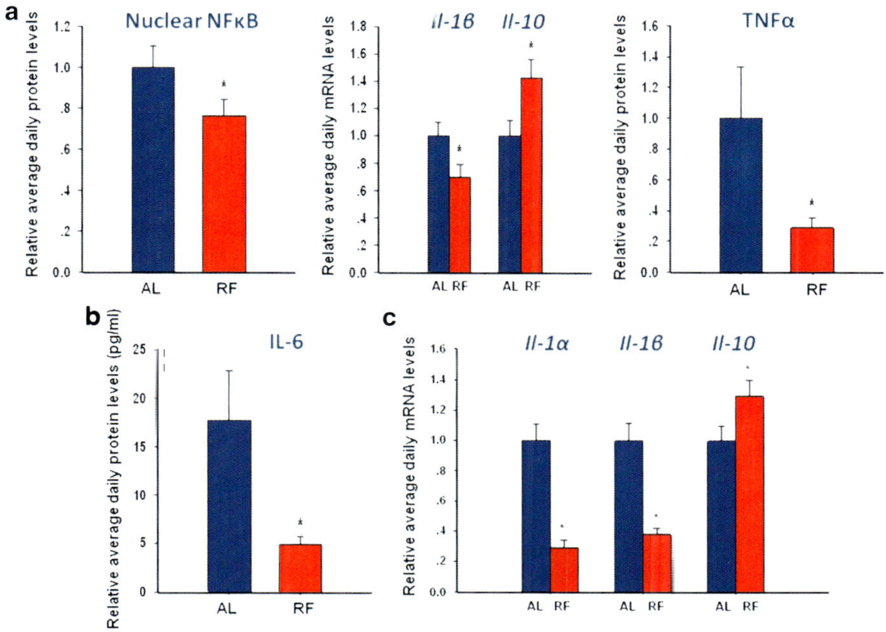

Fig. 5.2 Reduced levels of pro-inflammatory and increased levels of anti-inflammatory markers under RF in the liver (**a**), serum (**b**) and small intestine (**c**). *Nuclear NFκB* nuclear protein fraction of nuclear factor κB (pro-inflammatory), *Il-1α* interleukin 1α mRNA (pro-inflammatory), *Il-1β* interleukin 1β mRNA (pro-inflammatory), *Il-10* interleukin 10 mRNA (anti-inflammatory), *IL-6* interleukin 6 protein (pro-inflammatory), *TNFα* tumor necrosis factor α protein (pro-inflammatory)

[15, 55, 56, 64, 65, 67, 68]. Thus, RF uncouples the SCN from the periphery, suggesting that nutritional regulation of clock oscillators in peripheral tissues may play a direct role in coordinating metabolic oscillations [69]. Many physiological activities that are normally dictated by the SCN master clock, such as detoxification by hepatic P450 activity, body temperature, locomotor activity, and heart rate, are phase-shifted by RF to the time of food availability [63, 64, 70, 71]. As soon as food availability returns to be ad libitum, the SCN clock, whose phase remains unaffected, resets the peripheral oscillators [67]. It has recently been shown that long-term day-time RF can increase the amplitude of clock gene expression, increase expression of catabolic factors, and reduce the levels of disease markers leading to better health [72] (Fig. 5.2).

Because timed feeding is dominant in resetting circadian rhythms even in animals with lesioned SCN, it has been suggested that there is a food-entrainable oscillator. However, the location of this food-entrainable oscillator (FEO) has been elusive. Lesions in brain regions involved in feeding, such as the dorsomedial hypothalamic nucleus (DMH) [73–76], the brain stem parabrachial nuclei (PBN) [74, 77], and the core and shell regions of nucleus accumbens [78, 79], revealed that these nuclei may be involved in FEO output, but they cannot fully account for the oscillation [80]. Neither vagal signals nor leptin are critical for the entrainment [61, 81]. CLOCK [82] or BMAL1 [83] and other clock genes [84] have been shown not to be necessary for food anticipatory activity. However, it has recently been demonstrated that *Per2* mutant mice did not exhibit wheel-running food anticipation [85, 86]. Thus, how RF entrains circadian rhythms remains an extremely important topic for research.

Effect of Calorie Restriction (CR) on Circadian Rhythms

CR refers to a dietary regimen low in calories without malnutrition. CR restricts the amount of calories derived from carbohydrates, fats, or proteins to 60–75% of ad libitum-fed animals [87]. It has been documented that calorie restriction significantly extends the life span of rodents by up to 50% [88, 89]. In addition to the increase in life span, CR also delays the occurrence of age-associated pathophysiological changes, such as cancer, diabetes, kidney disease, and cataracts [89–92]. Theories on how CR modulates aging and longevity abound, but the exact mechanism is still unknown [89]. As opposed to RF, CR entrains the clock in the SCN [93–96], indicating that calorie reduction could affect the central oscillator. CR during the daytime affects the temporal organization of the SCN clockwork and circadian outputs in mice under light/dark cycle. In addition, CR affects responses of the circadian system to light, indicating that energy metabolism modulates gating of photic inputs in mammals [97]. These findings suggest that synchronization of peripheral oscillators during CR could be achieved directly due to the temporal eating, as has been reported for RF [64, 67, 68], or by synchronizing the SCN [93–95], which, in turn, sends humoral or neuronal signals to synchronize the peripheral tissues [98, 99].

Effect of Intermittent Fasting (IF) on Circadian Rhythms

During IF, food is available ad libitum every other day. IF-treated mice eat on the days they have access to food approximately twice as much as those having continuous access to food [100, 101]. Similarly to calorically restricted animals, IF-fed animals exhibit increased life span in comparison with the ad libitum-fed control [102] as well as improved glucose metabolism, cardio-protection, neuro-protection [100, 103–107], and increased resistance to cancer [101]. The IF-induced beneficial effects are thought to occur independently of the overall caloric intake, but the underlying mechanisms are still unknown. One suggested mechanism is stimulation of cellular stress pathways induced by the IF regimen [100, 108, 109]. Recently it has been shown that when food was introduced during the light period, mice exhibited almost arrhythmicity in clock gene expression in the liver (Fig. 5.3). Unlike daytime feeding, nighttime feeding yielded rhythms similar to those generated during ad libitum feeding [110]. The fact that IF can affect circadian rhythms differently

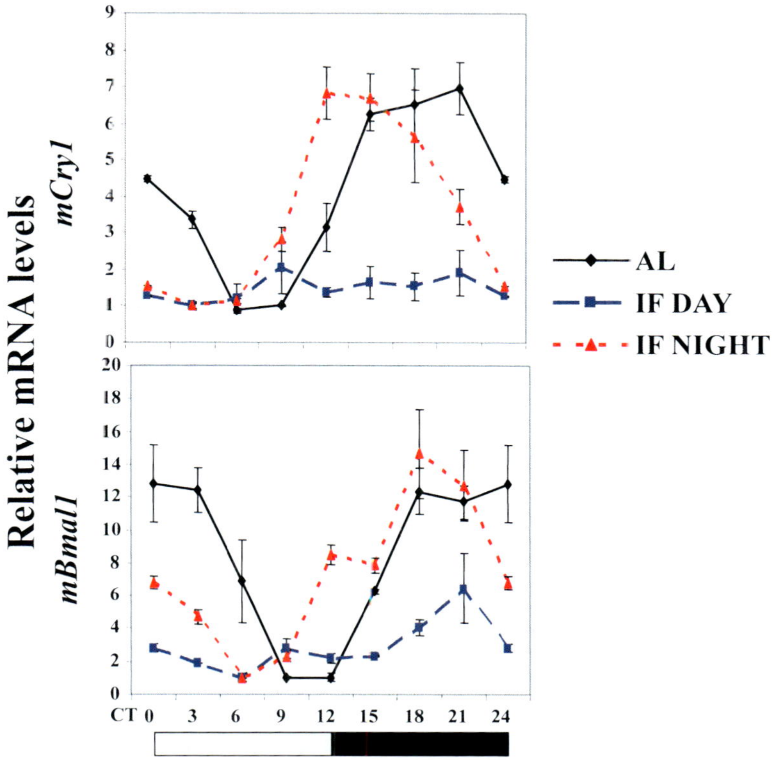

Fig. 5.3 Day-time IF abolishes circadian rhythms. mRNA expression levels of two mouse clock genes *mCry1* and *Bmal1* in the liver during ad libitum (AL) and day and night intermittent fasting (IF). The *gray* and *black bars* designate the subjective day (formerly the light period) and dark cycles, respectively. CT0 and CT12 represent the circadian times at which the lights would have been turned on and off, respectively, had the animals remained in light–dark. *CT* circadian time

depending on the timing of food availability suggests that this regimen affects the SCN clock, similarly to CR. SCN resetting by IF and CR could be involved in the health benefits conferred by these regimens [99].

Effect of High-Fat Diet on Circadian Rhythms

Few studies show that a high-fat diet leads to minimal effects on the rhythmic expression of clock genes in visceral adipose tissue and liver [111]. However, recent studies have shown that introduction of a high-fat diet to animals leads to rapid changes in both the period of locomotor activity in constant darkness and to increased food intake during the normal rest period under light–dark conditions [112]. These

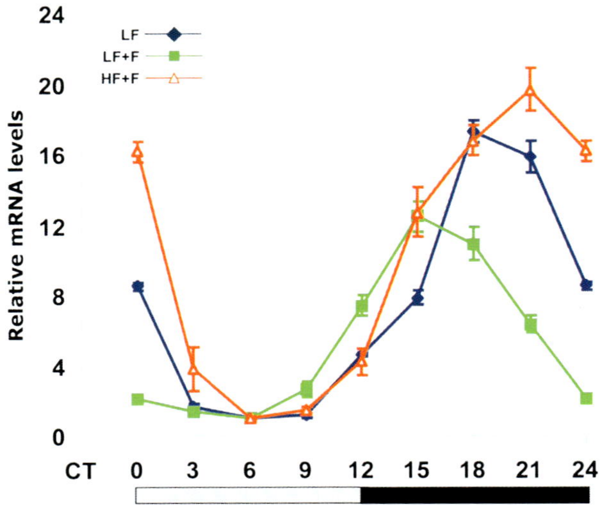

Fig. 5.4 Fasting induces phase advances while high-fat diet causes phase delays. mRNA expression levels of the clock gene *Bmal1* in the liver after fasting or high-fat diet. The *gray* and *black bars* designate the subjective day (formerly the light period) and dark cycles, respectively. CT0 and CT12 represent the circadian times at which the lights would have been turned on and off, respectively, had the animals remained in light–dark. *CT* circadian time, *LF* low-fat diet, *LF + F* low-fat diet + fasting, *HF + F* high-fat diet + fasting

changes in behavioral rhythmicity correlated with disrupted clock gene expression within hypothalamus, liver, and adipose tissue, as well as with altered cycling of hormones and nuclear hormone receptors involved in fuel utilization, such as leptin, thyroid stimulating hormone (TSH), and testosterone in mice, rats, and humans [112–117]. Furthermore, a high-fat diet modulates carbohydrate metabolism by amplifying circadian variation in glucose tolerance and insulin sensitivity [118].

In addition to the disruption of clock gene expression, high-fat diet induced a phase delay in clock and clock-controlled genes [116, 117] (Fig. 5.4). Recently, AMPK has been found to phosphorylate Ser-389 of CKIε, resulting in increased CKIε activity and degradation of PER2. PER2 degradation leads to a phase advance in the circadian expression pattern of clock genes in wild-type mice [119]. As the levels of AMPK decline under HF diet [116, 117], it is plausible that the changes seen in the expression phase of genes under HF diet are mediated by changes in AMPK levels. In addition to its effect on gene expression, high-fat feeding led to impaired adjustment to local time by light resetting, including slower rate of re-entrainment of behavioral and body temperature rhythms after "jet-lag" tests (6 h advanced light–dark cycle) and reduced phase-advance responses to light. These results correlated with reduction in c-FOS and phosphoERK expression in the SCN in response to light-induced phase shifts [120].

Summary Points

- Western lifestyle leads to high food consumption, inactivity during the active period, enhanced activity in the rest period, and shortened sleep period.
- Disrupted biological rhythms might lead to attenuated circadian feeding rhythms, disrupted metabolism, cancer proneness, and reduced life expectancy.
- Feeding time has the ability to reset bodily rhythms.
- Resetting the biological clock by food or feeding time may lead to better functionality of physiological systems, preventing metabolic disorders, promoting well-being, and extending life span.

References

1. Wyatt SB, Winters KP, Dubbert PM (2006) Overweight and obesity: prevalence, consequences, and causes of a growing public health problem. Am J Med Sci 331:166–174
2. Oishi K, Shirai H, Ishida N (2005) CLOCK is involved in the circadian transactivation of peroxisome-proliferator-activated receptor alpha (PPARalpha) in mice. Biochem J 386:575–581
3. Turek FW, Joshu C, Kohsaka A, Lin E, Ivanova G, McDearmon E, Laposky A, Losee-Olson S, Easton A, Jensen DR, Eckel RH, Takahashi JS, Bass J (2005) Obesity and metabolic syndrome in circadian *Clock* mutant mice. Science 308:1043–1045
4. Marcheva B, Ramsey KM, Buhr ED, Kobayashi Y, Su H, Ko CH, Ivanova G, Omura C, Mo S, Vitaterna MH, Lopez JP, Philipson LH, Bradfield CA, Crosby SD, JeBailey L, Wang X, Takahashi JS, Bass J (2010) Disruption of the clock components CLOCK and BMAL1 leads to hypoinsulinaemia and diabetes. Nature 466:627–631
5. Froy O (2010) Metabolism and circadian rhythms—implications for obesity. Endocr Rev 31:1–24
6. Arble DM, Bass J, Laposky AD, Vitaterna MH, Turek FW (2009) Circadian timing of food intake contributes to weight gain. Obesity (Silver Spring) 17:2100–2102
7. Reppert SM, Weaver DR (2002) Coordination of circadian timing in mammals. Nature 418:935–941
8. Penev PD, Kolker DE, Zee PC, Turek FW (1998) Chronic circadian desynchronization decreases the survival of animals with cardiomyopathic heart disease. Am J Physiol 275:H2334–H2337
9. Fu L, Pelicano H, Liu J, Huang P, Lee C (2002) The circadian gene *Period2* plays an important role in tumor suppression and DNA damage response *in vivo*. Cell 111:41–50
10. Filipski E, King VM, Li X, Granda TG, Mormont MC, Claustrat B, Hastings MH, Levi F (2003) Disruption of circadian coordination accelerates malignant growth in mice. Pathol Biol 51:216–219
11. Davis S, Mirick DK (2006) Circadian disruption, shift work and the risk of cancer: a summary of the evidence and studies in Seattle. Cancer Causes Control 17:539–545
12. Hurd MW, Ralph MR (1998) The significance of circadian organization for longevity in the golden hamster. J Biol Rhythms 13:430–436
13. Karasek M (2004) Melatonin, human aging, and age-related diseases. Exp Gerontol 39:1723–1729
14. Garaulet M, Madrid JA (2010) Chronobiological aspects of nutrition, metabolic syndrome and obesity. Adv Drug Deliv Rev 62:967–978
15. Hirota T, Fukada Y (2004) Resetting mechanism of central and peripheral circadian clocks in mammals. Zoolog Sci 21:359–368

16. Kohsaka A, Bass J (2007) A sense of time: how molecular clocks organize metabolism. Trends Endocrinol Metab 18:4–11
17. La Fleur SE, Kalsbeek A, Wortel J, Buijs RM (1999) A suprachiasmatic nucleus generated rhythm in basal glucose concentrations. J Neuroendocrinol 11:643–652
18. La Fleur SE (2003) Daily rhythms in glucose metabolism: suprachiasmatic nucleus output to peripheral tissue. J Neuroendocrinol 15:315–322
19. Davidson AJ, Castanon-Cervantes O, Stephan FK (2004) Daily oscillations in liver function: diurnal vs circadian rhythmicity. Liver Int 24:179–186
20. Ramsey KM, Marcheva B, Kohsaka A, Bass J (2007) The clockwork of metabolism. Annu Rev Nutr 27:219–240
21. Cailotto C, La Fleur SE, Van Heijningen C, Wortel J, Kalsbeek A, Feenstra M, Pevet P, Buijs RM (2005) The suprachiasmatic nucleus controls the daily variation of plasma glucose via the autonomic output to the liver: are the clock genes involved? Eur J Neurosci 22:2531–2540
22. Kalsbeek A, Ruiter M, La Fleur SE, Cailotto C, Kreier F, Buijs RM (2006) The hypothalamic clock and its control of glucose homeostasis. Prog Brain Res 153:283–307
23. Yamazaki S, Ishida Y, Inouye S (1994) Circadian rhythms of adenosine triphosphate contents in the suprachiasmatic nucleus, anterior hypothalamic area and caudate putamen of the rat—negative correlation with electrical activity. Brain Res 664:237–240
24. Ruiter M, La Fleur SE, van Heijningen C, van der Vliet J, Kalsbeek A, Buijs RM (2003) The daily rhythm in plasma glucagon concentrations in the rat is modulated by the biological clock and by feeding behavior. Diabetes 52:1709–1715
25. Ando H, Yanagihara H, Hayashi Y, Obi Y, Tsuruoka S, Takamura T, Kaneko S, Fujimura A (2005) Rhythmic messenger ribonucleic acid expression of clock genes and adipocytokines in mouse visceral adipose tissue. Endocrinology 146:5631–5636
26. De Boer SF, Van der Gugten J (1987) Daily variations in plasma noradrenaline, adrenaline and corticosterone concentrations in rats. Physiol Behav 40:323–328
27. Ahima RS, Prabakaran D, Flier JS (1998) Postnatal leptin surge and regulation of circadian rhythm of leptin by feeding. Implications for energy homeostasis and neuroendocrine function. J Clin Invest 101:1020–1027
28. Bodosi B, Gardi J, Hajdu I, Szentirmai E, Obal F Jr, Krueger JM (2004) Rhythms of ghrelin, leptin, and sleep in rats: effects of the normal diurnal cycle, restricted feeding, and sleep deprivation. Am J Physiol Regul Integr Comp Physiol 287:R1071–R1079
29. Kalra SP, Bagnasco M, Otukonyong EE, Dube MG, Kalra PS (2003) Rhythmic, reciprocal ghrelin and leptin signaling: new insight in the development of obesity. Regul Pept 111:1–11
30. Downs JL, Urbanski HF (2006) Aging-related sex-dependent loss of the circulating leptin 24-h rhythm in the rhesus monkey. J Endocrinol 190:117–127
31. Kalsbeek A, Fliers E, Romijn JA, La Fleur SE, Wortel J, Bakker O, Endert E, Buijs RM (2001) The suprachiasmatic nucleus generates the diurnal changes in plasma leptin levels. Endocrinology 142:2677–2685
32. Sukumaran S, Almon RR, DuBois DC, Jusko WJ (2010) Circadian rhythms in gene expression: Relationship to physiology, disease, drug disposition and drug action. Adv Drug Deliv Rev 62:904–917
33. Shen J, Tanida M, Niijima A, Nagai K (2007) In vivo effects of leptin on autonomic nerve activity and lipolysis in rats. Neurosci Lett 416:193–197
34. Licinio J (1998) Longitudinally sampled human plasma leptin and cortisol concentrations are inversely correlated. J Clin Endocrinol Metab 83:1042
35. Heptulla R, Smitten A, Teague B, Tamborlane WV, Ma YZ, Caprio S (2001) Temporal patterns of circulating leptin levels in lean and obese adolescents: relationships to insulin, growth hormone, and free fatty acids rhythmicity. J Clin Endocrinol Metab 86:90–96
36. Guan XM, Hess JF, Yu H, Hey PJ, van der Ploeg LH (1997) Differential expression of mRNA for leptin receptor isoforms in the rat brain. Mol Cell Endocrinol 133:1–7
37. Yi CX, van der Vliet J, Dai J, Yin G, Ru L, Buijs RM (2006) Ventromedial arcuate nucleus communicates peripheral metabolic information to the suprachiasmatic nucleus. Endocrinology 147:283–294

38. Zigman JM, Jones JE, Lee CE, Saper CB, Elmquist JK (2006) Expression of ghrelin receptor mRNA in the rat and the mouse brain. J Comp Neurol 494:528–548
39. Oishi K, Ohkura N, Wakabayashi M, Shirai H, Sato K, Matsuda J, Atsumi G, Ishida N (2006) CLOCK is involved in obesity-induced disordered fibrinolysis in ob/ob mice by regulating PAI-1 gene expression. J Thromb Haemost 4:1774–1780
40. Green CB, Takahashi JS, Bass J (2008) The meter of metabolism. Cell 134:728–742
41. Gorman MR (2003) Differential effects of multiple short day lengths on body weights of gonadectomized siberian hamsters. Physiol Biochem Zool 76:398–405
42. Morgan PJ, Ross AW, Mercer JG, Barrett P (2003) Photoperiodic programming of body weight through the neuroendocrine hypothalamus. J Endocrinol 177:27–34
43. Bocquier F, Bonnet M, Faulconnier Y, Guerre-Millo M, Martin P, Chilliard Y (1998) Effects of photoperiod and feeding level on perirenal adipose tissue metabolic activity and leptin synthesis in the ovariectomized ewe. Reprod Nutr Dev 38:489–498
44. Faulconnier Y, Bonnet M, Bocquier F, Leroux C, Chilliard Y (2001) Effects of photoperiod and feeding level on adipose tissue and muscle lipoprotein lipase activity and mRNA level in dry non-pregnant sheep. Br J Nutr 85:299–306
45. Di Lorenzo L, De Pergola G, Zocchetti C, L'Abbate N, Basso A, Pannacciulli N, Cignarelli M, Giorgino R, Soleo L (2003) Effect of shift work on body mass index: results of a study performed in 319 glucose-tolerant men working in a Southern Italian industry. Int J Obes Relat Metab Disord 27:1353–1358
46. Karlsson B, Knutsson A, Lindahl B (2001) Is there an association between shift work and having a metabolic syndrome? Results from a population based study of 27,485 people. Occup Environ Med 58:747–752
47. Karlsson BH, Knutsson AK, Lindahl BO, Alfredsson LS (2003) Metabolic disturbances in male workers with rotating three-shift work. Results of the WOLF study. Int Arch Occup Environ Health 76:424–430
48. Perfetto F, Tarquini R, Cornelissen G, Mello G, Tempestini A, Gaudiano P, Mancuso F, Halberg F (2004) Circadian phase difference of leptin in android versus gynoid obesity. Peptides 25:1297–1306
49. Zhang Y, Proenca R, Maffei M, Barone M, Leopold L, Friedman JM (1994) Positional cloning of the mouse obese gene and its human homologue. Nature 372:425–432
50. Montague CT, Farooqi IS, Whitehead JP, Soos MA, Rau H, Wareham NJ, Sewter CP, Digby JE, Mohammed SN, Hurst JA, Cheetham CH, Earley AR, Barnett AH, Prins JB, O'Rahilly S (1997) Congenital leptin deficiency is associated with severe early-onset obesity in humans. Nature 387:903–908
51. Gavrila A, Peng CK, Chan JL, Mietus JE, Goldberger AL, Mantzoros CS (2003) Diurnal and ultradian dynamics of serum adiponectin in healthy men: comparison with leptin, circulating soluble leptin receptor, and cortisol patterns. J Clin Endocrinol Metab 88:2838–2843
52. Yildiz BO, Suchard MA, Wong ML, McCann SM, Licinio J (2004) Alterations in the dynamics of circulating ghrelin, adiponectin, and leptin in human obesity. Proc Natl Acad Sci U S A 101:10434–10439
53. Puchalski SS, Green JN, Rasmussen DD (2003) Melatonin effect on rat body weight regulation in response to high-fat diet at middle age. Endocrine 21:163–167
54. Stephan FK (2002) The "other" circadian system: food as a *Zeitgeber*. J Biol Rhythms 17:284–292
55. Cassone VM, Stephan FK (2002) Central and peripheral regulation of feeding and nutrition by the mammalian circadian clock: implications for nutrition during manned space flight. Nutrition 18:814–819
56. Schibler U, Ripperger J, Brown SA (2003) Peripheral circadian oscillators in mammals: time and food. J Biol Rhythms 18:250–260
57. Honma KI, Honma S, Hiroshige T (1983) Critical role of food amount for prefeeding corticosterone peak in rats. Am J Physiol 245:R339–R344

102. Goodrick CL, Ingram DK, Reynolds MA, Freeman JR, Cider N (1990) Effects of intermittent feeding upon body weight and lifespan in inbred mice: interaction of genotype and age. Mech Ageing Dev 55:69–87
103. Contestabile A, Ciani E (2004) Dietary restriction differentially protects from neurodegeneration in animal models of excitotoxicity. Brain Res 1002:162–166
104. Mattson MP (2005) Energy intake, meal frequency, and health: a neurobiological perspective. Annu Rev Nutr 25:237–260
105. Sharma S, Kaur G (2005) Neuroprotective potential of dietary restriction against kainate-induced excitotoxicity in adult male Wistar rats. Brain Res Bull 67:482–491
106. Ahmet I, Wan R, Mattson MP, Lakatta EG, Talan M (2005) Cardioprotection by intermittent fasting in rats. Circulation 112:3115–3121
107. Mager DE, Wan R, Brown M, Cheng A, Wareski P, Abernethy DR, Mattson MP (2006) Caloric restriction and intermittent fasting alter spectral measures of heart rate and blood pressure variability in rats. FASEB J 20:631–637
108. Mattson MP, Duan W, Wan R, Guo Z (2004) Prophylactic activation of neuroprotective stress response pathways by dietary and behavioral manipulations. NeuroRx 1:111–116
109. Mattson MP (2008) Dietary factors, hormesis and health. Ageing Res Rev 7:43–48
110. Froy O, Chapnik N, Miskin R (2009) Effect of intermittent fasting on circadian rhythms in mice depends on feeding time. Mech Ageing Dev 130:154–160
111. Yanagihara H, Ando H, Hayashi Y, Obi Y, Fujimura A (2006) High-fat feeding exerts minimal effects on rhythmic mRNA expression of clock genes in mouse peripheral tissues. Chronobiol Int 23:905–914
112. Kohsaka A, Laposky AD, Ramsey KM, Estrada C, Joshu C, Kobayashi Y, Turek FW, Bass J (2007) High-fat diet disrupts behavioral and molecular circadian rhythms in mice. Cell Metab 6:414–421
113. Havel PJ, Townsend R, Chaump L, Teff K (1999) High-fat meals reduce 24-h circulating leptin concentrations in women. Diabetes 48:334–341
114. Cha MC, Chou CJ, Boozer CN (2000) High-fat diet feeding reduces the diurnal variation of plasma leptin concentration in rats. Metabolism 49:503–507
115. Cano P, Jimenez-Ortega V, Larrad A, Reyes Toso CF, Cardinali DP, Esquifino AI (2008) Effect of a high-fat diet on 24-h pattern of circulating levels of prolactin, luteinizing hormone, testosterone, corticosterone, thyroid-stimulating hormone and glucose, and pineal melatonin content, in rats. Endocrine 33:118–125
116. Barnea M, Madar Z, Froy O (2009) High-fat diet delays and fasting advances the circadian expression of adiponectin signaling components in mouse liver. Endocrinology 150:161–168
117. Barnea M, Madar Z, Froy O (2010) High-fat diet followed by fasting disrupts circadian expression of adiponectin signaling pathway in muscle and adipose tissue. Obesity (Silver Spring) 18:230–238
118. Rudic RD, McNamara P, Curtis AM, Boston RC, Panda S, Hogenesch JB, Fitzgerald GA (2004) BMAL1 and CLOCK, two essential components of the circadian clock, are involved in glucose homeostasis. PLoS Biol 2:e377
119. Um JH, Yang S, Yamazaki S, Kang H, Viollet B, Foretz M, Chung JH (2007) Activation of 5'-AMP-activated kinase with diabetes drug metformin induces casein kinase Iepsilon (CKIepsilon)-dependent degradation of clock protein mPER2. J Biol Chem 282:20794–20798
120. Mendoza J, Pevet P, Challet E (2008) High-fat feeding alters the clock synchronization to light. J Physiol 586:5901–5910

Chapter 6
Sleep, Energy Homeostasis and Metabolic Syndrome Alterations

Anthony H. Tsang, Jana Husse, and Henrik Oster

Abstract Though occupying up to 30% of our lifetime, the biological process of sleep retains many of its secrets. Most animals need to sleep regularly, but why this is essential for general well-being and life itself remains unknown. One important function of sleep lies in its regulation of metabolic homeostasis. In this chapter we describe the complex interactive relationship of sleep and metabolism and the impact of sleep loss and sleep disruption on the development of the metabolic syndrome. We show that the two processes are regulated by complementary and partially overlapping central circuits and both share a close connection with the circadian clock. In our modern societies sleep hygiene has long been neglected, but it becomes increasingly clear that healthy and sufficient sleep is an essential factor in maintaining a normal body weight and minimizing the risk of developing obesity-associated diseases such as type 2 diabetes and the metabolic syndrome.

Introduction

The need for regular sleep appears to be a universal property of all animals, suggesting a conserved mechanism [1]. Nevertheless, there are still major gaps in our understanding of sleep mechanisms and function. With an ongoing trend towards shorter sleep times and increasing interference of work demands with natural sleep

A.H. Tsang, M. Sc • J. Husse, Ph.D.
Circadian Rhythms Group, Max Planck Institute for Biophysical Chemistry, Am Fassberg 11, Göttingen 37077, Germany

H. Oster, Ph. D. (✉)
Circadian Rhythms Group, Max Planck Institute for Biophysical Chemistry, Am Fassberg 11, Göttingen 37077, Germany

Medical Department I, University of Lübeck, Lübeck, Germany
e-mail: Henrik.Oster@mpibpc.mpg.de

M. Garaulet and J.M. Ordovás (eds.), *Chronobiology and Obesity*,
DOI 10.1007/978-1-4614-5082-5_6, © Springer Science+Business Media New York 2013

rhythms in modern societies it becomes increasingly clear that "sound sleep" is more than a subjective indicator for well-being and quality of life. Sleep turns out to be an essential regulator of physiological and neurophysiological processes, in particular in the context of energy metabolism and against the background of an increasingly obese society.

Sleep Regulation and Function

Sleep is regulated by two processes termed "process C" and "process S" [2, 3]. "Process C" describes the regulation of the timing of vigilance states by the circadian clock, whereas "process S" is the homeostatic component of sleep. The interaction of these two allows a consolidated period of sleep, which—in humans and other diurnal species—is restricted to the night. "Process S," or sleep drive, increases with extended time spent awake. The circadian clock stimulates wakefulness especially during the afternoon when homeostatic sleep drive is high, and thus prevents us from going to bed too early. During the night, when sleep drive is reset, the circadian clock appears to stimulate sleep and to maintain it until the early morning [4]. An established physiological marker for sleep drive is delta power (EEG oscillations in the frequency range of 1–4 Hz). Delta power increases with increasing time spent awake [5] and the amount of delta power rebound during recovery from sleep deprivation is proportional to the amount of sleep loss [6]. In addition, delta power negatively correlates with sleep fragmentation [7] and the response to arousing stimuli [8], and can thus be used as a measure of sleep intensity. Neuroanatomically, "process C" can be located to the hypothalamic suprachiasmatic nucleus (SCN), whereas "process S" involves a dispersed but connected network of arousal-promoting and sleep-promoting centers [9]. Interestingly, the interaction between the homeostatic and the circadian process seems to be mutual. Firing rhythms in the SCN are influenced by behavioral states [10, 11] while genes, which have a function in the circadian clock mechanism also play a role in sleep homeostatic processes. Moreover, sleep deprivation affects clock gene activity in various tissues (reviewed in [12]).

The evolutionary relevant function of sleep is still an unsolved mystery. It was suggested that sleep facilitates restoration at cellular or systemic levels, that it may help to conserve energy and that it has an important role in neuronal plasticity and learning. The restorative theory of sleep predicts that sleep is necessary to restore vital functions by either replenishing molecules (such as ATP) that have been depleted during wakefulness or by degrading sleep-promoting substances (such as adenosine) that have accumulated during wakefulness [13, 14]. Additional support for a restorative function of sleep comes from studies showing that sleep deprivation negatively affects the immune response (reviewed in [15]). In addition, transcriptional changes associated with the wake state are most pronounced in genes involved in the cellular stress response (especially heat shock proteins), in immune responses and detoxification processes [16–18]. Zimmermann and colleagues suggested that energy-consuming processes such as protein synthesis decrease during extended wakefulness, which could lead to the depletion of arousal-promoting factors [19]. A second theory is based on energy conservation during sleep [20]. Human and animal studies show reduced energy expenditure during sleep as well as increased energy expenditure during sleep deprivation [21–23]. In addition, energy metabolism in the cortex is decreased during NREM sleep [24]. On the other hand, it has been shown that acute sleep deprivation can reduce energy expenditure in humans [25]. A third idea of why we sleep is to optimize neuronal performance by regulating neuronal plasticity. This hypothesis is built on the fact that sleep improves memory performance. If animals or humans are allowed to sleep after a learning task, memory consolidation is improved [26]. During sleep the hippocampus, a crucial brain area for learning and memory, appears to replay neuronal activation patterns that were also observed during a learning task in the preceding wake state, indicating that the reactivation of memory traces during sleep might be an important step in memory consolidation [27]. In addition, a substantial synchronization of firing patterns between hippocampal and cortical areas during sleep has been observed, which was interpreted as memory transfer from hippocampal short-term to long-term cortical storages [28]. Remarkably, experimentally boosting these neuronal replays by representing the before-learned stimulus during sleep enhances memory performance [29]. There is structural and molecular evidence for a net increase of synaptic strength during wakefulness and an overall decrease during sleep [30–33]. These data are also in accordance with the synaptic homeostasis theory, which postulates that sleep is necessary for a global synaptic downscaling to compensate for the net increase in synaptic strength during wakefulness (reviewed in [34]). In this context one could speculate that the brain needs a regular "offline" state in order to reorganize neuronal networks in the absence of sensory input, which would likely interfere with this process.

Metabolic Syndrome

In modern industrialized societies worldwide the prevalence of obesity is increasing at an alarming rate. Obesity itself is an important risk factor that predisposes individuals to several metabolic disorders and associated comorbidities such as type 2 diabetes (T2D), cardiovascular disease (CVD) and obstructive sleep apnea (OSA). These chronic metabolic diseases cast a heavy burden to suffering individuals, their families and the society as well. The term *metabolic syndrome* refers to a pathological condition that occurs as a combination of several metabolic abnormalities. According to the World Health Organization (WHO), it is defined by the presence of at least one of the following symptoms: impaired glucose regulation, diabetes mellitus and insulin resistance—associated with at least two of the following conditions: raised arterial pressure (≥160/90 mmHg), raised plasma triglycerides (≥150 mg dl^{-1}) and/or low HDL-cholesterol (<35 mg dl^{-1} for men; <39 mg dl^{-1} for women), development of central obesity (males: waist to hip ratio >0.90; females: waist to hip ratio >0.85) and/or BMI (>30 kg m^{-2}), and microalbuminuria (urinary albumin excretion rate ≥20 μg min^{-1}) or an albumin to creatinine ratio of more than 20 mg g^{-1} [35].

It has been proposed that sleep disturbance is an important contributing factor to the development of metabolic syndrome. A growing body of evidence, from well-controlled laboratory tests to large scale cross-sectional epidemiologic studies, has revealed a tight link between sleep disturbance and the pathogenesis of metabolic syndrome. The majority of these studies support a correlation between short sleep duration and an increased incidence of obesity and metabolic diseases in the past few decades (reviewed in [36]). While the prevalence of the metabolic syndrome has been increasing drastically over the last 30–40 years [37], compromised sleep quantity and quality have become more evident in the same period of time. For example, in the USA, comparing a large scale survey conducted in the 1960s by the American Cancer Society [38] and the more recent "Sleep in America" poll conducted in 2008 by the National Sleep Foundation [39] revealed a marked reduction in average sleep duration during workdays, from 8–9 h to 6–7 h. Another survey in the USA reported that the percentage of adults that sleep less than 6 h per night had increased by about 6% between 1985 and 2004 [40]. A similar trend was reported for European and Asian industrialized countries (reviewed in [41]). Moreover, in modern 24/7 societies, there is a growing population of night workers and rotating shift workers. This lifestyle leads to the misalignment between endogenous circadian rhythms and daily behavior. Such circadian desynchrony has been suggested to contribute to poor sleep quality and promote metabolic complications [42–44]. Therefore, a better understanding of the relationship between sleep hygiene and metabolic function is of urgent importance for clinicians to provide medical advice on sleep organization to patients with established metabolic disorders or individuals diagnosed with high risk to develop metabolic syndrome. The subsequent part of this chapter will discuss the relationship between sleep disturbances and major metabolic diseases.

Sleep and Glucose Metabolism

Blood glucose regulation is a tightly controlled physiological process. Both hypo- and hyperglycemia can have life-threatening consequences. The insulin signaling pathway plays a major role in regulating peripheral blood glucose levels. After a meal, blood glucose rises rapidly. Pancreatic beta cells respond to elevated glucose by releasing insulin to the blood stream, which promotes glucose uptake and utilization by insulin-sensitive organs such as liver, adipose tissues, and skeletal muscle. Insulin also inhibits glucogenic processes such as gluconeogenesis, lipolysis, and proteolysis. Insufficient insulin production from pancreatic beta cells or defects of the responsiveness of insulin-dependent target organs, a condition known as insulin resistance, causes glucose intolerance or diabetes [45]. In healthy individuals, glucose tolerance displays a circadian variation, which is higher in the evening than in the morning and reaches its nadir around midnight [46]. Interestingly, the timing of this midnight dip in glucose tolerance coincides with the occurrence of NREM sleep. A reduction in glucose utilization by the brain and the periphery during NREM sleep has been proposed as the cause of this rhythm [46]. One of the earliest experiments conducted by Kuhn et al. that linked sleep loss to compromised glucose metabolism involved subjecting healthy subjects to total sleep deprivation for 72–126 h [47]. These individuals showed a marked reduction in glucose tolerance throughout the testing period. While complete sleep deprivation is not common in humans, the more real life-related recurrent partial sleep restriction has also been shown to have an impact in glucose metabolism. A pilot laboratory study conducted by Spiegel et al. found that healthy young male subjects after being subjected to a sleep restriction protocol with a bedtime period reduced to 4 h per night for 6 days showed a 40% reduction in glucose tolerance [48]. Along the impaired glucose metabolism sleep-restricted subjects showed a reduced acute insulin response to glucose stimulation and reduced insulin sensitivity [48]. A more recent laboratory study revealed that even milder recurrent sleep restriction with 5.5 h of sleep per night negatively impacts on glucose tolerance and insulin sensitivity [49]. A negative effect of voluntary sleep curtailment on glucose metabolism has been reported in a number of large scale cross-sectional epidemiologic studies, carried out in several industrialized countries worldwide (reviewed in [41, 49]). These studies generally conclude that insufficient sleep predisposes individuals to glucose intolerance and increases the risk of developing diabetes, even after controlling for several covariants such as age, body mass index (BMI) and physical activity. Interestingly, some studies report a U-shaped relationship between sleep duration and glucose intolerance. Sleep of more than 8 h per night also associates with a higher risk of developing diabetes and may even impact negatively on life expectancy [41].

Not only is the duration of sleep important for proper glucose metabolism, but so is its quality. Slow Wave Sleep (SWS) has been proposed to be the most restorative among different stages of sleep in humans [50]. Reduced SWS is observed in elderly [51] and obese individuals [52, 53], and both groups are more vulnerable to develop T2D [54]. SWS can be experimentally suppressed and replaced with shallow NREM sleep by delivering acoustic stimuli while the total sleep duration is left unchanged. Tasali et al. employed this methodology to evaluate the importance of SWS in glucose metabolism of healthy individuals [55]. They reported SWS depletion of up to 90% which is comparable to what is seen in patients suffering from OSA (Obstructive sleep Apnea) [56]. After three nights of SWS deprivation, subjects showed significant glucose intolerance, together with reduced insulin sensitivity without a compensatory elevated insulin release [55]. This finding has been confirmed by a similar subsequent study which reported a marked reduction in glucose effectiveness and insulin sensitivity following SWS suppression by mechanical and acoustic stimuli for two nights [57]. Interestingly, diabetic patients seem to have less SWS than healthy individuals, suggesting a bidirectional interaction between glucose metabolism and sleep architecture [58]. In real life, compromised sleep quality can be seen most commonly in patients suffering from OSA. This breathing abnormality is characterized by the occurrence of repetitive episodes of upper airway blockades and, hence, intermittent hypoxia and sleep disturbance. OSA patients seldom wake

Despite the fact that most of the physiological consequences of glucocorticoids oppose those of insulin, it has also been reported that under chronic conditions, glucocorticoids may facilitate visceral fat accumulation (reviewed in [86]). Although the underlying mechanisms remain uncertain, a positive effect of glucocorticoids on de novo lipogenesis (DNL) and pre-adipocyte maturation may be involved [86]. Glucocorticoids work synergistically with insulin to promote DNL. Studies on adrenal-ectomized rats reveal that lipogenesis induced by refeeding after starvation is glucocorticoid-dependent [87]. The contribution of DNL to normal lipid homeostasis and in the context of metabolic syndrome has been highlighted by several recent studies in humans. Using a heavy water tracing technique, Strawford et al. showed that about 20% of TAGs in adipose tissues originate from DNL [88]. Moreover, hepatic DNL contributes to the development of fatty liver disease [89]. In addition, glucocorticoids promote the differentiation of pre-adipocytes into mature adipocytes. In pre-adipocyte cell cultures, the standard differentiation protocol typically includes addition of the glucocorticoid analog dexamethasone and insulin [90, 91]. Glucocorticoids may thus contribute to regulate fat mass by this differentiation promoting effect [92]. Taken together, sleep loss may lead to lipid dysregulation via excessive activation of lipolysis and subsequent ectopic lipid redistribution, by altering DNL, and by promoting pre-adipocyte differentiation.

Sleep and Peripheral Appetite Hormones

Epidemiological evidence suggests a negative correlation between sleep duration/quality and obesity. However, since most of the acute physiological effects of sleep disturbance are catabolic, they are unlikely to be the direct causal factors contributing to the observed obesity phenotype. Thus, the generally suspected culprit is a sleep loss-induced increase in food intake. A reduced amount of sleep at the same time means that the daily active phase is prolonged, which allows for more opportunities for eating, particularly, snacks [93]. This is especially true for modern 24/7 societies where food availability is not restrained by the time of day. Moreover, such increase in food intake may further contribute to the observed alterations in peripheral hormonal signaling.

Besides serving as a major energy reservoir of the organism, white adipose tissue is also a prominent endocrine organ which secretes a panel of circulating hormones and cytokines, known as adipo(cyto)kines that modulate a multitude of physiological functions [94]. The currently best characterized adipokine is leptin, a satiety hormone whose release is positively regulated by insulin-induced glucose uptake and overall fat mass [95, 96], both of which indicate a nutritionally rich status of the body. Leptin blood levels increase after meals. The hunger hormone ghrelin, which is secreted by the stomach, essentially works in an opposite way to leptin with release being suppressed after meals but increased during fasting. The blood concentrations of both hormones are higher during sleep compared to the wake state. Nevertheless, during the second half of the night, blood ghrelin levels are declining while leptin concentrations continue to rise, indicating that specific sleep stages might have differential effects of these two counteracting hormones [48, 72].

Sleep loss has been found to perturb the balance of leptin–ghrelin signaling. Using a partial sleep deprivation paradigm Spiegel et al. observed that sleep loss results in reduced plasma leptin levels while ghrelin levels are increased [48, 97]. An elevated plasma ghrelin-to-leptin ratio has well been correlated to self-reported appetite and hunger ratings, with preferences towards carbohydrate-rich meals [97]. Taken together, sleep loss seems to provoke hunger via perturbing the homeostatic balance between leptin and ghrelin release. As the postprandial increase of leptin [98]/decrease of ghrelin [99] has been shown to be promoted by insulin, sleep loss-related insulin resistance may contribute to ghrelin/leptin effects of sleep deprivation. Recently, the blood levels of a number of other blood-borne nutrient-related hormones such as Peptide YY (PYY) and nicotinamide phosphoribosyltransferase (Nampt/visfatin/PBEF) have also been reported to be affected by sleep [100, 101], but their contributions to the regulation of appetite and energy expenditure remain largely unclear.

Sleep and Cardiovascular Function

A final hallmark of the metabolic syndrome are cardiovascular disorders (CVDs), particularly high blood pressure. CVDs may be the most life-threatening consequence among the various symptoms described for metabolic syndrome. A number of epidemiologic studies establish a close association between sleep hygiene and cardiovascular function [102–104]. Several potential interacting mechanisms that link sleep disturbance to cardiovascular disease have been discussed above. Obesity and ectopic arterial lipid accumulation are likely long-term causes of CVDs [105]. The chronic over-activation of sympathetic tone after sleep loss may be an important contributor to raised blood pressure [106]. Moreover, elevated low-grade chronic inflammation is now appreciated to participate in all stages of the pathogenesis of many CVDs [70]. In addition, OSA represents another interesting link between CVD, obesity and sleep disturbances. OSA is a risk factor for several CVDs such as myocardial infarction and stroke independent of obesity [62]. Potential mechanisms include elevated sympathetic tone [62], oxidative stress elicited by recurrent hypoxic conditions [107], and blood vessel damage caused by loud snoring [108]. In this context it would be interesting to investigate if the CVD-predisposing effect of OSA is also mediated by its characteristic fragmenting effect on sleep architecture.

Sleep Anatomy

During recent years it has become increasingly clear that—like sleep—appetite and body weight regulation are centrally controlled processes [109, 110]. Thus, shared neuronal circuits may explain why in so many aspects sleep and metabolic regulation

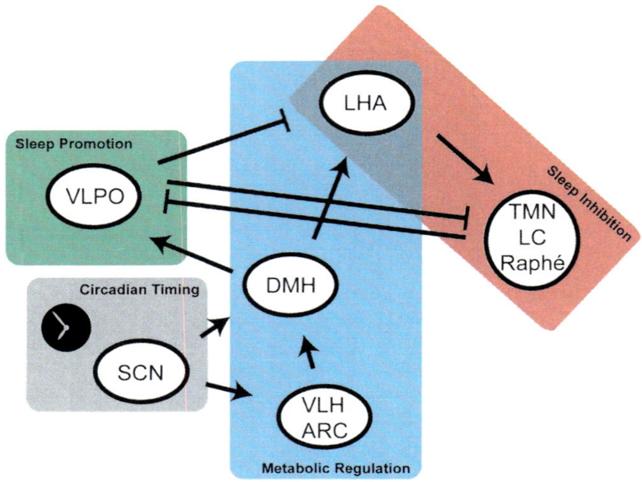

Fig. 6.1 Schematic illustration of mammalian sleep and metabolism regulating neural circuitries. Sleep regulatory neurons are organized in a mutually inhibitory network. The major wake-promoting centers are located in the upper brainstem. They project cholinergic, monoaminergic, and glutamatergic fibers to various brain regions including the cerebral cortex, thalamus, and hypothalamus. The main sleep-promoting center is located in the VLPO, in the anterior hypothalamus. It projects inhibitory GABAergic fibers to the brainstem arousal centers. The ascending arousal pathways are also capable of inhibiting the VLPO. This mutually inhibitory connection allows for rapid transitions between sleep and wake states. The mediobasal hypothalamus is the integrating center coordinating the homeostasis of metabolism and sleep. The wake-promoting orexinergic neurons in the LHA integrate the metabolic information and relay it to the brainstem wake-promoting centers. LHA neurons also directly innervate the cerebral cortex. The SCN provides the circadian input to regulate metabolism and sleep via direct connection to several nuclei in the mediobasal hypothalamus. *Abbreviations*: *ARC* arcuate nucleus, *DMH* dorsomedial hypothalamic nucleus, *LHA* lateral hypothalamic area, *SCN* suprachiasmatic nucleus, *VLPO* ventrolateral pre-optic nucleus, *TMN* tuberomammillary nucleus, *LC* locus coeruleus

are tightly interconnected (Fig. 6.1). Neuroanatomically, sleep depends on a network of widespread, but highly connected brain areas (reviewed in [9]). Among the wake-promoting centers, some fire only during wakefulness ("wake-on"), whereas others are active during wakefulness as well as during REM sleep ("REM-on"), indicating that the latter support the cortical activation observed during REM sleep. Monoaminergic cell groups in the midbrain and the *pons* are specifically active during wakefulness and target the lateral hypothalamus, the basal forebrain and cortical areas. These include the noradrenergic *locus coeruleus* (LC), the serotonergic dorsal and median raphé nuclei (DRN, MRN), dopaminergic neurons bordering the raphé nuclei, and histaminergic neurons of the tuberomammillary nucleus (TMN) [111–115]. In addition, glutamatergic neurons in the *pons* (in the parabrachial nucleus (PBN) and the pre-coeruleus area) stimulate arousal by innervation of hypothalamic, forebrain and cortical regions [116, 117]. The forebrain also contains arousal-promoting neuronal networks amongst which the orexin (also known as hypocretin) expressing neurons in the lateral hypothalamus have been studied extensively. Orexigenic neurons are most active during wakefulness, in particular during active exploration of the environment [118, 119]. They receive input from arousal-promoting areas (e.g., LC, DRN, PBN) as well as from brain areas involved in reward processing (e.g., amygdala and ventral tegmental area). Orexigenic neurons project to the *cortex*, the brainstem and forebrain areas, most densely to the TMN and the LC [9]. Genetic impairment of orexigenic signaling results in narcoleptic phenotypes in mice, dogs and humans [116, 117, 120]. Narcolepsy is characterized by a severe increase in the number of transitions between different vigilance states: human patients as well as mice lacking orexin tend to fall asleep more often and show more fragmented sleep in addition to more transitions between REM and NREM sleep [9]. Patients suffered from narcolepsy often also suffer from cataplexy (i.e., a sudden loss of muscle tone in response to emotional stimuli). Such patients have a greatly reduced number of orexigenic neurons in the hypothalamus and it has been proposed that at least in some individuals this reduction might be due to an autoimmune response [120–123].

NREM sleep-promoting areas are mainly found in the hypothalamus: GABAergic or galaninergic neurons in the ventrolateral preoptic area (VLPO) are selectively active during NREM sleep and innervate many of the above described arousal-promoting areas [124]. Interestingly, loss of VLPO neurons in human patients as well as in animals results in profound insomnia with an up to 50% reduction of sleep [125]. In close proximity to the VLPO another population of sleep-active neurons is found in the median preoptic nucleus (MnPO) [126, 127]. Interestingly, these neurons become active already before actual sleep onset, indicating a potential role in integrating accumulating sleep pressure. REM sleep-promoting neurons are located in the cholinergic pedunculopontine and the laterodorsal tegemental nuclei (PPT and LDT) and mainly innervate thalamic relay nuclei and the basal forebrain [128, 129]. REM-on neurons are found in the sub-laterodorsal nucleus (SLD), the precoeruleus region (PC), and the medial parabrachial nucleus (MPB). The SLD is involved in regulating atonia during REM sleep by activating inhibitory neurons in medulla and spinal cord, whereas PC and MPB neurons stimulate forebrain pathways resulting in the high-frequency EEG traces during REM sleep [117, 130, 131]. A remarkable feature of wake-, REM-, and NREM-promoting nuclei is that they seem to mutually inhibit each other. This lead to the so-called "flip-flop switch model" of state transitions suggesting that mutual inhibition leads to fast and complete transitions between different vigilance states, which can, for example, ensure rapid transitions to wakefulness in case of threat [9].

Hypothalamic Control of Sleep and Metabolism

The hypothalamus serves as an integrating center involved in controlling "homeostatic" behaviors and physiology including feeding, body temperature, breathing, cardiovascular function, metabolism, and sleep. The hypothalamic appetite regulating center is comprised of mutually inhibiting circuits within the arcuate nucleus (ARC). The ARC is the major targeting site for both leptin and ghrelin. It is composed of appetite-promoting agouti-related protein (AgRP)/neuropeptide Y (NPY) expressing neurons and appetite-suppressing cocaine and amphetamine-related transcript (CART)/pro-opiomelanocortin (POMC) expressing neurons. Leptin inhibits orexigenic AgRP/NPY neurons while activating anorexigenic CART/POMC neurons; ghrelin on the other hand activates AgRP/NPY neurons, but inhibits CART/POMC neurons. Moreover, both types of neurons heavily project to and mutually inhibit each other [109].

In a similar way, neurons in the lateral hypothalamus (LHA) are organized in local mutually inhibitory networks. In the LHA, wake-promoting neurons are glutamatergic and express orexin while sleep promoting neurons are GABAergic and express melanin-concentrating hormone (MCH) [132]. Multiple lines of evidence indicate that these two types of sleep regulating neurons are at the same time tightly involved in metabolic regulation. From a conceptual point of view it is easily conceivable that the homeostasis of sleep and metabolism has to be regulated in a highly cooperative manner as sleeping and feeding are two mutually exclusive behaviors. Both orexin and MCH neurons are regulated by metabolic signals of the body. Orexin neurons are the downstream targets of ARC AgRP/NPY neurons. NPY inhibits orexin neurons via Y1 and Y2 receptor subtypes [133, 134]. Interestingly, ghrelin, whose effect on NPY neurons is activating, also stimulates orexin neurons directly [135]. Along the same line elevation of several satiety signals including leptin, glucose, and free fatty acids/TAGs has been shown to inhibit orexinergic neuronal activity [135, 136], suggesting that peripheral satiety signals may suppress the wake-promoting function of the LHA. Sleep-promoting MCH neurons also respond to the change of metabolic status. Fasting has been reported to stimulate MCH neurons [137, 138]. MCH neurons are regulated in a largely opposite manner than orexin neurons. Acetylcholine and norepinephrine stimulate orexin neurons but have been shown to suppress MCH neurons [139], whereas on the one hand glucose inhibits orexin neurons, and on the other hand it activates MCH neurons [140].

Both orexin and MCH neurons play an active role in metabolic regulation. Intracerebroventricular (i.c.v.) injection of orexin increases food consumption [141]. Genetic depletion of orexin signaling in mice results in hypophagic and narcoleptic phenotypes, paralleled with reduced food-seeking behavior and energy expenditure [22, 135]. At the cellular level, orexin positively regulates AgRP/NPY, but inhibits CART/POMC neurons [142]. Sleep-deprived rodents show an activated orexin system [143]. MCH is also orexigenic as it has been shown by early studies that i.c.v. injection can stimulate feeding behavior in rodents [137, 138]. Moreover, elevated hypothalamic MCH signaling has anabolic effects such as reduced energy expenditure, increased lipid accumulation in adipose tissues and weight gain [144, 145]. Consistently, genetic deletion of either MCH or its receptors leads to weight loss and reduced fat mass and causes resistance to high caloric diet-induced obesity [146, 147]. The integrative picture on how these complex signaling pathways within the hypothalamus organize themselves to achieve a remarkably stable homeostatic regulation of sleep and metabolism and how they interact with other brain areas, e.g., those involved in cognitive and reward processes (see below), is still largely incomplete. However, the existence of bidirectional interactions between hypothalamic sleep and metabolic regulating centers is evident.

It has been shown that the main increase in caloric intake in sleep-deprived individuals stems from snacks while homeostatic feeding is relatively unaffected [93]. Therefore, it has been proposed that hedonic circuits may be activated by sleep disturbance. The reward system in the brain is known to be involved in sleep regulation [148]. It is suppressed during sleep; whereas when the reward system is activated, the activity of several sleep centers in the brain is inhibited [149]. The reward system involves mesolimbic brain structures including the ventral tegmental area (VTA) and the *nucleus accumbens* (NAc). These structures interact functionally with the hypothalamic sleep and metabolic regulating centers. For example, orexin has been shown to activate VTA dopaminergic neurons and promote dopamine release in the NAc, the target area of the VTA [150, 151]. NPY neurons also project to the NAc where they may modify local dopamine signaling [152, 153]. Leptin has been shown to decrease while ghrelin increases the activity of VTA dopaminergic neurons *in vitro*. *In vivo*, when applied directly to the VTA, leptin reduces while ghrelin and orexin promote food intake [154–156]. Thus, metabolic and sleep regulating signaling plays an active role in modifying the reward system and in this way sleep disturbances are capable of promoting hedonic behaviors [148, 157]. Interestingly, the addiction to certain activities, for example, video game playing or gambling has been reported to contribute to voluntary sleep loss [158–160], highlighting the bidirectional link between sleep and reward systems in the brain.

Conclusion

Epidemiological studies strongly suggest that sleep disturbance and sleep curtailment are important risk factors for the development of metabolic syndrome. Sleep loss has profound effects on metabolic regulation, but experimental data are still sparse on how chronic sleep disruption may impact on energy homeostasis and which of the involved processes are critical for the emergence of metabolic syndrome. An intriguing finding is that central circuits of sleep and appetite regulation are closely entangled. The analysis of the functional interaction between sleep and metabolic regulatory neurons might be key for a better understanding of the metabolic syndrome pathology in the context of shift work and sleep-affecting disorders and critically help clinicians in providing medical advice on lifestyle interventions to patients.

Summary Points

- Amount and quality of sleep have strong effects on metabolic homeostasis.
- A lack of sleep negatively impacts on various aspects of energy metabolism including glucose and lipid utilization, adipokine regulation and cardiovascular function.

- Sleep and metabolic regulatory centers in the brain are tightly connected.
- In the hypothalamus, orexigenic neurons control both arousal and appetite.
- Circadian rhythm disruption is an important risk factor for both metabolism and sleep disorders.

Acknowledgments A.H.T. is supported by an excellence fellowship of the GGNB graduate program of the University of Göttingen. J.H. is supported by the Max Planck Society. H.O. is an Emmy Noether fellow of the German Research Foundation (DFG) and a Lichtenberg fellow of the Volkswagen Foundation.

References

1. Zimmerman JE, Naidoo N, Raizen DM, Pack AI (2008) Conservation of sleep: insights from non-mammalian model systems. Trends Neurosci 31(7):371–376
2. Borbely AA (1982) A two process model of sleep regulation. Hum Neurobiol 1(3):195–204
3. Daan S, Beersma DG, Borbely AA (1984) Timing of human sleep: recovery process gated by a circadian pacemaker. Am J Physiol 246(2 Pt 2):R161–R183
4. Mistlberger RE (2005) Circadian regulation of sleep in mammals: role of the suprachiasmatic nucleus. Brain Res Rev 49(3):429–454
5. Dijk DJ, Beersma DG, Daan S (1987) EEG power density during nap sleep: reflection of an hourglass measuring the duration of prior wakefulness. J Biol Rhythms 2(3):207–219
6. Tobler I, Borbely AA (1986) Sleep EEG in the rat as a function of prior waking. Electroencephalogr Clin Neurophysiol 64(1):74–76
7. Franken P, Dijk DJ, Tobler I, Borbely AA (1991) Sleep deprivation in rats: effects on EEG power spectra, vigilance states, and cortical temperature. Am J Physiol 261(1 Pt 2):R198–R208
8. Neckelmann D, Ursin R (1993) Sleep stages and EEG power spectrum in relation to acoustical stimulus arousal threshold in the rat. Sleep 16(5):467–477
9. Saper CB, Fuller PM, Pedersen NP, Lu J, Scammell TE (2010) Sleep state switching. Neuron 68(6):1023–1042
10. Deboer T, Detari L, Meijer JH (2007) Long term effects of sleep deprivation on the mammalian circadian pacemaker. Sleep 30(3):257–262
11. Deboer T, Vansteensel MJ, Detari L, Meijer JH (2003) Sleep states alter activity of suprachiasmatic nucleus neurons. Nat Neurosci 6(10):1086–1090
12. Landgraf D, Shostak A, Oster H (2011) Clock genes and sleep. Pflugers Arch 463(1):3–14
13. Porkka-Heiskanen T, Strecker RE, Thakkar M, Bjørkum AA, Greene RW, McCarley RW (1997) Adenosine: a mediator of the sleep-inducing effects of prolonged wakefulness. Science 276(5316):1265–1268
14. Dworak M, McCarley RW, Kim T, Kalinchuk AV, Basheer R (2010) Sleep and brain energy levels: ATP changes during sleep. J Neurosci 30(26):9007–9016
15. Bollinger T, Bollinger A, Oster H, Solbach W (2010) Sleep, immunity, and circadian clocks: a mechanistic model. Gerontology 56(6):574–580
16. Andretic R, Franken P, Tafti M (2008) Genetics of sleep. Annu Rev Genet 42:361–388
17. Thompson CL, Wisor JP, Lee CK, Pathak SD, Gerashchenko D, Smith KA et al (2010) Molecular and anatomical signatures of sleep deprivation in the mouse brain. Front Neurosci 4:165
18. Cirelli C, Gutierrez CM, Tononi G (2004) Extensive and divergent effects of sleep and wakefulness on brain gene expression. Neuron 41(1):35–43
19. Zimmerman JE, Rizzo W, Shockley KR, Raizen DM, Naidoo N, Mackiewicz M et al (2006) Multiple mechanisms limit the duration of wakefulness in Drosophila brain. Physiol Genomics 27(3):337–350

20. Berger RJ, Phillips NH (1995) Energy conservation and sleep. Behav Brain Res 69(1–2):65–73
21. Jung CM, Melanson EL, Frydendall EJ, Perreault L, Eckel RH, Wright KP (2011) Energy expenditure during sleep, sleep deprivation and sleep following sleep deprivation in adult humans. J Physiol 589(1):235–244
22. Zhang S, Zeitzer JM, Sakurai T, Nishino S, Mignot E (2007) Sleep/wake fragmentation disrupts metabolism in a mouse model of narcolepsy. J Physiol 581(Pt 2):649–663
23. Rechtschaffen A, Bergmann BM (1995) Sleep deprivation in the rat by the disk-over-water method. Behav Brain Res 69(1–2):55–63
24. Braun AR, Balkin TJ, Wesenten NJ, Carson RE, Varga M, Baldwin P et al (1997) Regional cerebral blood flow throughout the sleep-wake cycle. An H2(15)O PET study. Brain 120(Pt 7):1173–1197
25. Benedict C, Hallschmid M, Lassen A, Mahnke C, Schultes B, Birgir Schioth H et al (2011) Acute sleep deprivation reduces energy expenditure in healthy men. Am J Clin Nutr 93(6):1229–1236
26. Diekelmann S, Born J (2010) The memory function of sleep. Nat Rev Neurosci 11(2):114–126
27. Wilson MA, McNaughton BL (1994) Reactivation of hippocampal ensemble memories during sleep. Science 265(5172):676–679
28. Ji D, Wilson MA (2007) Coordinated memory replay in the visual cortex and hippocampus during sleep. Nat Neurosci 10(1):100–107
29. Rasch B, Buchel C, Gais S, Born J (2007) Odor cues during slow-wave sleep prompt declarative memory consolidation. Science 315(5817):1426–1429
30. Bushey D, Tononi G, Cirelli C (2011) Sleep and synaptic homeostasis: structural evidence in Drosophila. Science 332(6037):1576–1581
31. Appelbaum L, Wang G, Yokogawa T, Skariah GM, Smith SJ, Mourrain P et al (2010) Circadian and homeostatic regulation of structural synaptic plasticity in hypocretin neurons. Neuron 68(1):87–98
32. Gilestro GF, Tononi G, Cirelli C (2009) Widespread changes in synaptic markers as a function of sleep and wakefulness in Drosophila. Science 324(5923):109–112
33. Maret S, Faraguna U, Nelson AB, Cirelli C, Tononi G (2011) Sleep and waking modulate spine turnover in the adolescent mouse cortex. Nat Neurosci 14(11):1418–1420
34. Wang G, Grone B, Colas D, Appelbaum L, Mourrain P (2011) Synaptic plasticity in sleep: learning, homeostasis and disease. Trends Neurosci 34(9):452–463
35. Alberti KG, Zimmet PZ (1998) Definition, diagnosis and classification of diabetes mellitus and its complications. Part 1: diagnosis and classification of diabetes mellitus provisional report of a WHO consultation. Diabet Med 15(7):539–553
36. Beccuti G, Pannain S (2011) Sleep and obesity. Curr Opin Clin Nutr Metab Care 14(4):402–412
37. Finucane MM, Stevens GA, Cowan MJ, Danaei G, Lin JK, Paciorek CJ et al (2011) National, regional, and global trends in body-mass index since 1980: systematic analysis of health examination surveys and epidemiological studies with 960 country-years and 9.1 million participants. Lancet 377(9765):557–567
38. Kripke DF, Simons RN, Garfinkel L, Hammond EC (1979) Short and long sleep and sleeping pills. Is increased mortality associated? Arch Gen Psychiatry 36(1):103–116
39. National Sleep Foundation. "Sleep in America" poll (2008) Available from http://www.sleepfoundation.org/sites/default/files/2008%20POLL%20SOF.PDF
40. National Center for Health Statistics (2005) QuickStats: percentage of adults who reported an average of ≤6 h of sleep per 24-hour period, by sex and age group - United States, 1984 and 2004. MMWR Morb Mortal Wkly Rep. 2005;54:933
41. Knutson KL, Van Cauter E (2008) Associations between sleep loss and increased risk of obesity and diabetes. Ann N Y Acad Sci 1129:287–304
42. Akerstedt T (2003) Shift work and disturbed sleep/wakefulness. Occup Med (Lond) 53(2):89–94
43. Ohayon MM, Lemoine P, Arnaud-Briant V, Dreyfus M (2002) Prevalence and consequences of sleep disorders in a shift worker population. J Psychosom Res 53(1):577–583

44. Scheer FA, Hilton MF, Mantzoros CS, Shea SA (2009) Adverse metabolic and cardiovascular consequences of circadian misalignment. Proc Natl Acad Sci USA 106(11):4453–4458
45. Nussey S, Whitehead S (2001) The endocrine pancreas. In: Endocrinology: an integrated approach. Oxford: BIOS Scientific Publishers
46. Van Cauter E, Blackman JD, Roland D, Spire JP, Refetoff S, Polonsky KS (1991) Modulation of glucose regulation and insulin secretion by circadian rhythmicity and sleep. J Clin Invest 88(3):934–942
47. Kuhn E, Brodan V, Brodanova M, Rysanek K (1969) Metabolic reflection of sleep deprivation. Act Nerv Super (Praha) 11(3):165–174
48. Spiegel K, Leproult R, Van Cauter E (1999) Impact of sleep debt on metabolic and endocrine function. Lancet 354(9188):1435–1439
49. Nedeltcheva AV, Kessler L, Imperial J, Penev PD (2009) Exposure to recurrent sleep restriction in the setting of high caloric intake and physical inactivity results in increased insulin resistance and reduced glucose tolerance. J Clin Endocrinol Metab 94(9):3242–3250
50. Maquet P (1995) Sleep function(s) and cerebral metabolism. Behav Brain Res 69(1–2):75–83
51. Van Cauter E, Leproult R, Plat L (2000) Age-related changes in slow wave sleep and REM sleep and relationship with growth hormone and cortisol levels in healthy men. JAMA 284(7):861–868
52. Resta O, Foschino Barbaro MP, Bonfitto P, Giliberti T, Depalo A, Pannacciulli N et al (2003) Low sleep quality and daytime sleepiness in obese patients without obstructive sleep apnoea syndrome. J Intern Med 253(5):536–543
53. Vgontzas AN, Tan TL, Bixler EO, Martin LF, Shubert D, Kales A (1994) Sleep apnea and sleep disruption in obese patients. Arch Intern Med 154(15):1705–1711
54. Mokdad AH, Bowman BA, Ford ES, Vinicor F, Marks JS, Koplan JP (2001) The continuing epidemics of obesity and diabetes in the United States. JAMA 286(10):1195–1200
55. Tasali E, Leproult R, Ehrmann DA, Van Cauter E (2008) Slow-wave sleep and the risk of type 2 diabetes in humans. Proc Natl Acad Sci U S A 105(3):1044–1049
56. Bourke SC, Gibson GJ (2002) Sleep and breathing in neuromuscular disease. Eur Respir J 19(6):1194–1201
57. Stamatakis KA, Punjabi NM (2010) Effects of sleep fragmentation on glucose metabolism in normal subjects. Chest 137(1):95–101
58. Pallayova M, Donic V, Gresova S, Peregrim I, Tomori Z (2010) Do differences in sleep architecture exist between persons with type 2 diabetes and nondiabetic controls? J Diabetes Sci Technol 4(2):344–352
59. Caples SM, Gami AS, Somers VK (2005) Obstructive sleep apnea. Ann Intern Med 142(3):187–197
60. Young T, Palta M, Dempsey J, Skatrud J, Weber S, Badr S (1993) The occurrence of sleep-disordered breathing among middle-aged adults. N Engl J Med 328(17):1230–1235
61. Tasali E, Mokhlesi B, Van Cauter E (2008) Obstructive sleep apnea and type 2 diabetes: interacting epidemics. Chest 133(2):496–506
62. Corsonello A, Antonelli Incalzi R, Pistelli R, Pedone C, Bustacchini S, Lattanzio F (2011) Comorbidities of chronic obstructive pulmonary disease. Curr Opin Pulm Med 17(Suppl 1):S21–S28
63. Mendoza J, Pevet P, Challet E (2008) High-fat feeding alters the clock synchronization to light. J Physiol 586(Pt 24):5901–5910
64. Kohsaka A, Laposky AD, Ramsey KM, Estrada C, Joshu C, Kobayashi Y et al (2007) High-fat diet disrupts behavioral and molecular circadian rhythms in mice. Cell Metab 6(5):414–421
65. Benedict C, Hallschmid M, Lassen A, Mahnke C, Schultes B, Schioth HB et al (2011) Acute sleep deprivation reduces energy expenditure in healthy men. Am J Clin Nutr 93(6):1229–1236
66. Klingenberg L, Sjodin A, Holmback U, Astrup A, Chaput JP (2012) Short sleep duration and its association with energy metabolism. Obes Rev 13(7):565–577
67. Thomas M, Sing H, Belenky G, Holcomb H, Mayberg H, Dannals R et al (2000) Neural basis of alertness and cognitive performance impairments during sleepiness. I. Effects of 24 h of sleep deprivation on waking human regional brain activity. J Sleep Res 9(4):335–352

68. Mullington JM, Simpson NS, Meier-Ewert HK, Haack M (2010) Sleep loss and inflammation. Best Pract Res Clin Endocrinol Metab 24(5):775–784
69. Imeri L, Opp MR (2009) How (and why) the immune system makes us sleep. Nat Rev Neurosci 10(3):199–210
70. Lumeng CN, Saltiel AR (2011) Inflammatory links between obesity and metabolic disease. J Clin Invest 121(6):2111–2117
71. Shoelson SE, Lee J, Goldfine AB (2006) Inflammation and insulin resistance. J Clin Invest 116(7):1793–1801
72. Spiegel K, Leproult R, Colecchia EF, L'Hermite-Balериaux M, Nie Z, Copinschi G et al (2000) Adaptation of the 24-h growth hormone profile to a state of sleep debt. Am J Physiol Regul Integr Comp Physiol 279(3):R874–R883
73. Spiegel K, Knutson K, Leproult R, Tasali E, Van Cauter E (2005) Sleep loss: a novel risk factor for insulin resistance and Type 2 diabetes. J Appl Physiol 99(5):2008–2019
74. Darmon P, Dadoun F, Boullu-Ciocca S, Grino M, Alessi MC, Dutour A (2006) Insulin resistance induced by hydrocortisone is increased in patients with abdominal obesity. Am J Physiol Endocrinol Metab 291(5):E995–E1002
75. Seino S, Shibasaki T, Minami K (2011) Dynamics of insulin secretion and the clinical implications for obesity and diabetes. J Clin Invest 121(6):2118–2125
76. Sun K, Kusminski CM, Scherer PE (2011) Adipose tissue remodeling and obesity. J Clin Invest 121(6):2094–2101
77. Broussard J, Brady MJ (2010) The impact of sleep disturbances on adipocyte function and lipid metabolism. Best Pract Res Clin Endocrinol Metab 24(5):763–773
78. Boden G (1997) Role of fatty acids in the pathogenesis of insulin resistance and NIDDM. Diabetes 46(1):3–10
79. Itani SI, Ruderman NB, Schmieder F, Boden G (2002) Lipid-induced insulin resistance in human muscle is associated with changes in diacylglycerol, protein kinase C, and IkappaB-alpha. Diabetes 51(7):2005–2011
80. Cunha DA, Ladriere L, Ortis F, Igoillo-Esteve M, Gurzov EN, Lupi R et al (2009) Glucagon-like peptide-1 agonists protect pancreatic beta-cells from lipotoxic endoplasmic reticulum stress through upregulation of BiP and JunB. Diabetes 58(12):2851–2862
81. Boden G, Lebed B, Schatz M, Homko C, Lemieux S (2001) Effects of acute changes of plasma free fatty acids on intramyocellular fat content and insulin resistance in healthy subjects. Diabetes 50(7):1612–1617
82. Hucking K, Hamilton-Wessler M, Ellmerer M, Bergman RN (2003) Burst-like control of lipolysis by the sympathetic nervous system in vivo. J Clin Invest 111(2):257–264
83. Coppack SW, Jensen MD, Miles JM (1994) In vivo regulation of lipolysis in humans. J Lipid Res 35(2):177–193
84. Macfarlane DP, Forbes S, Walker BR (2008) Glucocorticoids and fatty acid metabolism in humans: fuelling fat redistribution in the metabolic syndrome. J Endocrinol 197(2):189–204
85. Baxter JD, Forsham PH (1972) Tissue effects of glucocorticoids. Am J Med 53(5):573–589
86. Roberge C, Carpentier AC, Langlois MF, Baillargeon JP, Ardilouze JL, Maheux P et al (2007) Adrenocortical dysregulation as a major player in insulin resistance and onset of obesity. Am J Physiol Endocrinol Metab 293(6):E1465–E1478
87. Williams BH, Berdanier CD (1982) Effects of diet composition and adrenalectomy on the lipogenic responses of rats to starvation-refeeding. J Nutr 112(3):534–541
88. Strawford A, Antelo F, Christiansen M, Hellerstein MK (2004) Adipose tissue triglyceride turnover, de novo lipogenesis, and cell proliferation in humans measured with $2H_2O$. Am J Physiol Endocrinol Metab 286(4):E577–E588
89. Donnelly KL, Smith CI, Schwarzenberg SJ, Jessurun J, Boldt MD, Parks EJ (2005) Sources of fatty acids stored in liver and secreted via lipoproteins in patients with nonalcoholic fatty liver disease. J Clin Invest 115(5):1343–1351
90. Hauner H, Schmid P, Pfeiffer EF (1987) Glucocorticoids and insulin promote the differentiation of human adipocyte precursor cells into fat cells. J Clin Endocrinol Metab 64(4):832–835

91. Halvorsen YD, Bond A, Sen A, Franklin DM, Lea-Currie YR, Sujkowski D et al (2001) Thiazolidinediones and glucocorticoids synergistically induce differentiation of human adipose tissue stromal cells: biochemical, cellular, and molecular analysis. Metabolism 50(4):407–413
92. Tchoukalova YD, Fitch M, Rogers PM, Covington JD, Henagan TM, Ye J et al (2012) In vivo adipogenesis in rats measured by cell kinetics in adipocytes and plastic-adherent stroma-vascular cells in response to high-fat diet and thiazolidinedione. Diabetes 61(1):137–144
93. Nedeltcheva AV, Kilkus JM, Imperial J, Kasza K, Schoeller DA, Penev PD (2009) Sleep curtailment is accompanied by increased intake of calories from snacks. Am J Clin Nutr 89(1):126–133
94. Rondinone CM (2006) Adipocyte-derived hormones, cytokines, and mediators. Endocrine 29(1):81–90
95. Leibel RL (2002) The role of leptin in the control of body weight. Nutr Rev 60(10 Pt 2):S15–S19, discussion S68–S84, 5–7
96. Friedman JM (2002) The function of leptin in nutrition, weight, and physiology. Nutr Rev 60(10 Pt 2):S1–S14, discussion S68–S84, 5–7
97. Spiegel K, Tasali E, Penev P, Van Cauter E (2004) Brief communication: sleep curtailment in healthy young men is associated with decreased leptin levels, elevated ghrelin levels, and increased hunger and appetite. Ann Intern Med 141(11):846–850
98. Mizuno T, Bergen H, Kleopoulos S, Bauman WA, Mobbs CV (1996) Effects of nutritional status and aging on leptin gene expression in mice: importance of glucose. Horm Metab Res 28(12):679–684
99. Saad MF, Bernaba B, Hwu CM, Jinagouda S, Fahmi S, Kogosov E et al (2002) Insulin regulates plasma ghrelin concentration. J Clin Endocrinol Metab 87(8):3997–4000
100. Magee CA, Melanson EL, Frydendall EJ, Huang XF, Iverson DC, Caputi P (2009) Acute sleep restriction alters neurocrine hormones and appetite in healthy male adults. Sleep Biol Rhythms 7:125–127
101. Benedict C, Shostak A, Lange T, Brooks SJ, Schioth HB, Schultes B et al (2012) Diurnal rhythm of circulating nicotinamide phosphoribosyltransferase (Nampt/visfatin/PBEF): impact of sleep loss and relation to glucose metabolism. J Clin Endocrinol Metab 97(2):E218–E222
102. Gottlieb DJ, Redline S, Nieto FJ, Baldwin CM, Newman AB, Resnick HE et al (2006) Association of usual sleep duration with hypertension: the Sleep Heart Health Study. Sleep 29(8):1009–1014
103. Gangwisch JE, Heymsfield SB, Boden-Albala B, Buijs RM, Kreier F, Pickering TG et al (2006) Short sleep duration as a risk factor for hypertension: analyses of the first National Health and Nutrition Examination Survey. Hypertension 47(5):833–839
104. Kotani K, Saiga K, Sakane N, Mu H, Kurozawa Y (2008) Sleep status and blood pressure in a healthy normotensive female population. Int J Cardiol 125(3):425–427
105. Greenberg AS, Coleman RA, Kraemer FB, McManaman JL, Obin MS, Puri V et al (2011) The role of lipid droplets in metabolic disease in rodents and humans. J Clin Invest 121(6):2102–2110
106. Spiegel K, Leproult R, L'Hermite-Baleriaux M, Copinschi G, Penev PD, Van Cauter E (2004) Leptin levels are dependent on sleep duration: relationships with sympathovagal balance, carbohydrate regulation, cortisol, and thyrotropin. J Clin Endocrinol Metab 89(11):5762–5771
107. Lavie L (2003) Obstructive sleep apnoea syndrome–an oxidative stress disorder. Sleep Med Rev 7(1):35–51
108. Lee SA, Amis TC, Byth K, Larcos G, Kairaitis K, Robinson TD et al (2008) Heavy snoring as a cause of carotid artery atherosclerosis. Sleep 31(9):1207–1213
109. Sanchez-Lasheras C, Konner AC, Bruning JC (2010) Integrative neurobiology of energy homeostasis-neurocircuits, signals and mediators. Front Neuroendocrinol 31(1):4–15
110. Kleinridders A, Konner AC, Bruning JC (2009) CNS-targets in control of energy and glucose homeostasis. Curr Opin Pharmacol 9(6):794–804

111. Aston-Jones G, Bloom FE (1981) Activity of norepinephrine-containing locus coeruleus neurons in behaving rats anticipates fluctuations in the sleep-waking cycle. J Neurosci 1(8):876–886
112. Dahlstroem A, Fuxe K (1964) Evidence for the existence of monoamine-containing neurons in the central nervous system. I. Demonstration of monoamines in the cell bodies of brain stem neurons. Acta Physiol Scand Suppl. suppl 232:1–55
113. Kocsis B, Varga V, Dahan L, Sik A (2006) Serotonergic neuron diversity: identification of raphe neurons with discharges time-locked to the hippocampal theta rhythm. Proc Natl Acad Sci USA 103(4):1059–1064
114. Lu J, Jhou TC, Saper CB (2006) Identification of wake-active dopaminergic neurons in the ventral periaqueductal gray matter. J Neurosci 26(1):193–202
115. Steininger TL, Alam MN, Gong H, Szymusiak R, McGinty D (1999) Sleep-waking discharge of neurons in the posterior lateral hypothalamus of the albino rat. Brain Res 840(1–2):138–147
116. Saper CB, Loewy AD (1980) Efferent connections of the parabrachial nucleus in the rat. Brain Res 197(2):291–317
117. Lu J, Sherman D, Devor M, Saper CB (2006) A putative flip-flop switch for control of REM sleep. Nature 441(7093):589–594
118. Lee MG, Hassani OK, Jones BE (2005) Discharge of identified orexin/hypocretin neurons across the sleep-waking cycle. J Neurosci 25(28):6716–6720
119. Mileykovskiy BY, Kiyashchenko LI, Siegel JM (2005) Behavioral correlates of activity in identified hypocretin/orexin neurons. Neuron 46(5):787–798
120. Peyron C, Faraco J, Rogers W, Ripley B, Overeem S, Charnay Y et al (2000) A mutation in a case of early onset narcolepsy and a generalized absence of hypocretin peptides in human narcoleptic brains. Nat Med 6(9):991–997
121. Thannickal TC, Moore RY, Nienhuis R, Ramanathan L, Gulyani S, Aldrich M et al (2000) Reduced number of hypocretin neurons in human narcolepsy. Neuron 27(3):469–474
122. Cvetkovic-Lopes V, Bayer L, Dorsaz S, Maret S, Pradervand S, Dauvilliers Y et al (2010) Elevated Tribbles homolog 2-specific antibody levels in narcolepsy patients. J Clin Invest 120(3):713–719
123. Kawashima M, Lin L, Tanaka S, Jennum P, Knudsen S, Nevsimalova S et al (2010) Anti-Tribbles homolog 2 (TRIB2) autoantibodies in narcolepsy are associated with recent onset of cataplexy. Sleep 33(7):869–874
124. Sherin JE, Shiromani PJ, McCarley RW, Saper CB (1996) Activation of ventrolateral preoptic neurons during sleep. Science 271(5246):216–219
125. Lu J, Greco MA, Shiromani P, Saper CB (2000) Effect of lesions of the ventrolateral preoptic nucleus on NREM and REM sleep. J Neurosci 20(10):3830–3842
126. Takahashi K, Lin JS, Sakai K (2009) Characterization and mapping of sleep-waking specific neurons in the basal forebrain and preoptic hypothalamus in mice. Neuroscience 161(1):269–292
127. Suntsova N, Szymusiak R, Alam MN, Guzman-Marin R, McGinty D (2002) Sleep-waking discharge patterns of median preoptic nucleus neurons in rats. J Physiol 543(Pt 2):665–677
128. el Mansari M, Sakai K, Jouvet M (1989) Unitary characteristics of presumptive cholinergic tegmental neurons during the sleep-waking cycle in freely moving cats. Exp Brain Res 76(3):519–29
129. Sakai K, Jouvet M (1980) Brain stem PGO-on cells projecting directly to the cat dorsal lateral geniculate nucleus. Brain Res 194(2):500–505
130. Luppi PH, Gervasoni D, Boissard R, Verret L, Goutagny R, Peyron C et al (2004) Brainstem structures responsible for paradoxical sleep onset and maintenance. Arch Ital Biol 142(4):397–411
131. Vetrivelan R, Fuller PM, Tong Q, Lu J (2009) Medullary circuitry regulating rapid eye movement sleep and motor atonia. J Neurosci 29(29):9361–9369
132. Bonnavion P, de Lecea L (2010) Hypocretins in the control of sleep and wakefulness. Curr Neurol Neurosci Rep 10(3):174–179

133. Fu LY, Acuna-Goycolea C, van den Pol AN (2004) Neuropeptide Y inhibits hypocretin/orexin neurons by multiple presynaptic and postsynaptic mechanisms: tonic depression of the hypothalamic arousal system. J Neurosci 24(40):8741–8751
134. Acuna-Goycolea C, Tamamaki N, Yanagawa Y, Obata K, van den Pol AN (2005) Mechanisms of neuropeptide Y, peptide YY, and pancreatic polypeptide inhibition of identified green fluorescent protein-expressing GABA neurons in the hypothalamic neuroendocrine arcuate nucleus. J Neurosci 25(32):7406–7419
135. Yamanaka A, Beuckmann CT, Willie JT, Hara J, Tsujino N, Mieda M et al (2003) Hypothalamic orexin neurons regulate arousal according to energy balance in mice. Neuron 38(5):701–713
136. Jordan SD, Konner AC, Bruning JC (2010) Sensing the fuels: glucose and lipid signaling in the CNS controlling energy homeostasis. Cell Mol Life Sci 67(19):3255–3273
137. Qu D, Ludwig DS, Gammeltoft S, Piper M, Pelleymounter MA, Cullen MJ et al (1996) A role for melanin-concentrating hormone in the central regulation of feeding behaviour. Nature 380(6571):243–247
138. Rossi M, Choi SJ, O'Shea D, Miyoshi T, Ghatei MA, Bloom SR (1997) Melanin-concentrating hormone acutely stimulates feeding, but chronic administration has no effect on body weight. Endocrinology 138(1):351–355
139. Bayer L, Eggermann E, Serafin M, Grivel J, Machard D, Muhlethaler M et al (2005) Opposite effects of noradrenaline and acetylcholine upon hypocretin/orexin versus melanin concentrating hormone neurons in rat hypothalamic slices. Neuroscience 130(4):807–811
140. Burdakov D, Gerasimenko O, Verkhratsky A (2005) Physiological changes in glucose differentially modulate the excitability of hypothalamic melanin-concentrating hormone and orexin neurons in situ. J Neurosci 25(9):2429–2433
141. Sakurai T, Amemiya A, Ishii M, Matsuzaki I, Chemelli RM, Tanaka H, et al (1998) Orexins and orexin receptors: a family of hypothalamic neuropeptides and G protein-coupled receptors that regulate feeding behavior. Cell;92(5):1 page following 696
142. Muroya S, Funahashi H, Yamanaka A, Kohno D, Uramura K, Nambu T et al (2004) Orexins (hypocretins) directly interact with neuropeptide Y, POMC and glucose-responsive neurons to regulate Ca2+ signaling in a reciprocal manner to leptin: orexigenic neuronal pathways in the mediobasal hypothalamus. Eur J Neurosci 19(6):1524–1534
143. Martins PJ, Marques MS, Tufik S, D'Almeida V (2010) Orexin activation precedes increased NPY expression, hyperphagia, and metabolic changes in response to sleep deprivation. Am J Physiol Endocrinol Metab 298(3):E726–E734
144. Ludwig DS, Tritos NA, Mastaitis JW, Kulkarni R, Kokkotou E, Elmquist J et al (2001) Melanin-concentrating hormone overexpression in transgenic mice leads to obesity and insulin resistance. J Clin Invest 107(3):379–386
145. Della-Zuana O, Presse F, Ortola C, Duhault J, Nahon JL, Levens N (2002) Acute and chronic administration of melanin-concentrating hormone enhances food intake and body weight in Wistar and Sprague–Dawley rats. Int J Obes Relat Metab Disord 26(10):1289–1295
146. Shimada M, Tritos NA, Lowell BB, Flier JS, Maratos-Flier E (1998) Mice lacking melanin-concentrating hormone are hypophagic and lean. Nature 396(6712):670–674
147. Bjursell M, Gerdin AK, Ploj K, Svensson D, Svensson L, Oscarsson J et al (2006) Melanin-concentrating hormone receptor 1 deficiency increases insulin sensitivity in obese leptin-deficient mice without affecting body weight. Diabetes 55(3):725–733
148. Hasler BP, Smith LJ, Cousins JC, Bootzin RR (2012) Circadian rhythms, sleep, and substance abuse. Sleep Med Rev 16(1):67–81
149. Albrecht U (2012) Timing to perfection: the biology of central and peripheral circadian clocks. Neuron 74(2):246–260
150. Korotkova TM, Sergeeva OA, Eriksson KS, Haas HL, Brown RE (2003) Excitation of ventral tegmental area dopaminergic and nondopaminergic neurons by orexins/hypocretins. J Neurosci 23(1):7–11
151. Narita M, Nagumo Y, Miyatake M, Ikegami D, Kurahashi K, Suzuki T (2007) Implication of protein kinase C in the orexin-induced elevation of extracellular dopamine levels and its rewarding effect. Eur J Neurosci 25(5):1537–1545

152. Kelley AE (2004) Ventral striatal control of appetitive motivation: role in ingestive behavior and reward-related learning. Neurosci Biobehav Rev 27(8):755–776
153. Gray TS, Morley JE (1986) Neuropeptide Y: anatomical distribution and possible function in mammalian nervous system. Life Sci 38(5):389–401
154. Muschamp JW, Dominguez JM, Sato SM, Shen RY, Hull EM (2007) A role for hypocretin (orexin) in male sexual behavior. J Neurosci 27(11):2837–2845
155. Naleid AM, Grace MK, Cummings DE, Levine AS (2005) Ghrelin induces feeding in the mesolimbic reward pathway between the ventral tegmental area and the nucleus accumbens. Peptides 26(11):2274–2279
156. Hommel JD, Trinko R, Sears RM, Georgescu D, Liu ZW, Gao XB et al (2006) Leptin receptor signaling in midbrain dopamine neurons regulates feeding. Neuron 51(6):801–810
157. Gujar N, Yoo SS, Hu P, Walker MP (2011) Sleep deprivation amplifies reactivity of brain reward networks, biasing the appraisal of positive emotional experiences. J Neurosci 31(12):4466–4474
158. Kevitiyagala D, Finlay F, Baverstock A (2011) Question 1. What is the impact of computer games on sleep in children? Arch Dis Child 96(9):894–895
159. Parhami I, Siani A, Rosenthal RJ, Fong TW (2012) Pathological gambling, problem gambling and sleep complaints: an analysis of the national comorbidity survey: replication (NCS-R). J Gambl Stud [Epub ahead of print]
160. Parhami I, Siani A, Rosenthal RJ, Lin S, Collard M, Fong TW (2012) Sleep and gambling severity in a community sample of gamblers. J Addict Dis 31(1):67–79

Chapter 7
Increased Risk of Diabetes due to Obesity: Does Chronodisruption Play a Role?

Susanne E. la Fleur and Andries Kalsbeek

Abstract The prevalence of obesity and type 2 diabetes mellitus (T2DM) has risen to epidemic proportions. The pathophysiology of T2DM is complex and involves insulin resistance, pancreatic β-cell dysfunction and visceral adiposity. Although it has been known for quite some time that a disruption of biological rhythms (as happens with shift work) increases the risk of developing obesity, insulin resistance and T2DM, more recent genomic evidence has further spiked the interest for the involvement of circadian rhythms (and their disruption) in the development of diabetes. In this chapter, we will start with an overview of the way in which glucose metabolism and the basal rhythm in plasma glucose concentrations and insulin sensitivity are regulated, after which we will discuss how a disruption of daily rhythms or a disruption of clock elements, may contribute to the development of insulin resistance.

Glucose Metabolism: An Introduction

Blood glucose is the most important energy source for the central nervous system. Since glycogen is neither stored nor produced in the central nervous system (CNS), plasma glucose concentrations must be maintained at a sufficiently high level for the organism to survive. It is therefore not surprising that the brain has several regulatory

S.E. la Fleur, Ph.D. (✉)
Department of Endocrinology and Metabolism, Academic Medical Center, University of Amsterdam, Meibergdreef 9, Amsterdam, 1105 AZ, The Netherlands
e-mail: s.e.lafleur@amc.uva.nl

A. Kalsbeek, Ph.D.
Department of Endocrinology and Metabolism, Academic Medical Center, University of Amsterdam, Meibergdreef 9, Amsterdam, 1105 AZ, The Netherlands

Hypothalamic Integration Mechanisms, Netherlands Institute for Neuroscience, 1105 BA, Amsterdam, The Netherlands

M. Garaulet and J.M. Ordovás (eds.), *Chronobiology and Obesity*,
DOI 10.1007/978-1-4614-5082-5_7,

pathways to ensure these sufficiently high levels. This was already apparent to Claude Bernard in 1855, when he performed his famous experiments showing that when the floor of the fourth ventricle of the brain was pricked with a needle, this resulted in a rise in sugar in the urine. What he also proposed was that the sugar was secreted from the liver into the blood as a consequence of "internal secretion." Although his experiments were far from precise and the exact site of his pricking has never been established [1, 2], the conclusion drawn from these and following experiments, i.e., that blood glucose levels are under the influence of the central nervous system, is now widely accepted. Almost a century after Bernard's experiments the term *homeostasis* was introduced by Walter B. Cannon. He extended Bernard's concept of "milieu interieur" and hypothesized that the maintenance of a relatively constant internal environment is of crucial importance for an organism [3]. This constant internal environment is maintained by neuronal and endocrine systems that activate compensating mechanisms when they detect disturbances. For the homeostatic regulation of blood glucose levels this means secretion of hormones such as insulin and glucagon and activation of the autonomic nervous system in terms of innervating the peripheral organs (i.e., liver, pancreas, and adrenals) (Box 7.1). Although this concept of homeostasis is key to glucose metabolism, plasma glucose concentrations are not kept at a constant level, but show a clear daily rhythm with the highest levels found prior to the onset of the activity period.

Box 7.1 Hormones that influence plasma glucose concentrations

Plasma glucose concentrations are the net result of glucose output and glucose uptake. Certain organs, hormones and metabolites are important to keep glucose concentrations within physiological boundaries. During feeding, glucose is taken up from the gut, and in times of fasting, glucose production in the liver is stimulated, resulting in an increased output of glucose from the liver to the circulation. Glucose produced by the liver is the result of glycogen breakdown (glycogenolysis) and the forming of new glucose from lactate and amino acids (gluconeogenesis). A number of hormones regulate glucose production and glucose uptake:

- *Insulin* is released by the pancreatic β-cell when glucose concentrations are high. It stimulates the uptake of glucose in muscle and adipose tissue. At the same time, at the level of the liver, insulin suppresses the production of glucose. Both processes thus result in a lowering of plasma glucose concentrations. However, when insulin sensitivity is impaired, glucose concentrations in plasma will rise due to a diminished uptake, but also due to an increased glucose production by the liver as a consequence of the lesser suppression by insulin.

- *Glucagon* is released from the pancreatic α-cells, when plasma glucose concentrations drop. Glucagon acts on the liver to stimulate glucose production, but probably also reduces glucose uptake in liver and muscle.
- *Corticosterone* (or cortisol in humans) is released by the adrenal cortex in response to stress via ACTH. Corticosterone acts on the liver to increase glucose production, especially through gluconeogenesis, to ensure enough glucose within the blood stream during times of fasting.
- *Adrenalin* (also known as epinephrine) is released by the adrenal medulla in response to stress via sympathetic stimulation. Its effects are similar to those of corticosterone but are more rapid because of its primary focus on glycogenolysis.
- *Noradrenalin* (also known as norepinephrine) is also released from the adrenal medulla, but also by sympathetic nerve endings. Noradrenaline stimulates glucose production by its release from sympathetic nerve endings in the liver, but at the same time also stimulates the release of glucagon and inhibits the release of insulin by its release from sympathetic nerve endings in the pancreatic α- and β-cells.

In addition to their action within peripheral organs, all these hormones also have receptors in the central nervous system and thus may also affect glucose metabolism via the hypothalamus.

Changes in plasma glucose concentrations result from food intake, hepatic glucose production (via glycogenolysis or gluconeogenesis), and/or changes in glucose uptake by tissues such as the brain, muscle, liver and adipose tissue. An increased glucose uptake (i.e., higher glucose tolerance) may be the result of various processes, such as higher amounts of insulin being secreted from the pancreatic β-cells after a meal, an increased insulin sensitivity of peripheral tissues, or an increase in (non)-insulin-mediated glucose uptake via the recruitment of glucose transporters in exercising muscles (Box 7.2). Most, if not all, processes that regulate basal glucose concentrations show daily rhythms.

Box 7.2

To accommodate the effects of food intake and glucose utilization, plasma glucose concentrations are contained within strict boundaries by various internal mechanisms. Plasma glucose concentrations rise with increased hepatic glucose production via glycogenolysis or gluconeogenesis:

Glycogenolysis: Conversion of glycogen polymers to glucose monomers. Glucagon and epinephrine stimulate glycogenolysis.

(continued)

Box 7.2 (continued)

Gluconeogenesis: Biosynthesis of "new" glucose from non-glucose precursors, such as lactate and amino acids. Corticosterone stimulates gluconeogenesis.

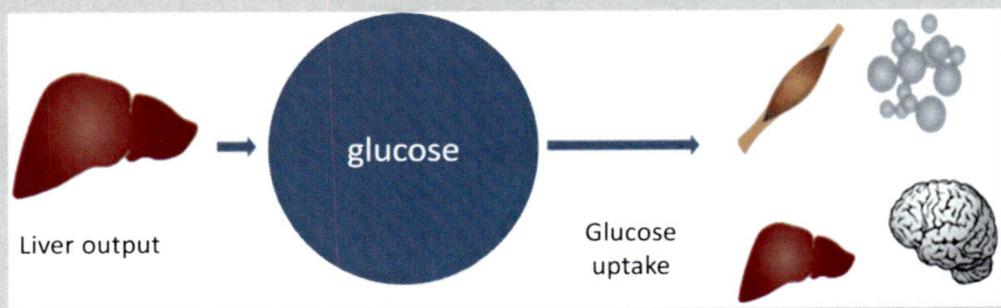

Plasma glucose concentrations may also vary as a result of changes in glucose uptake in tissues such as the brain, muscle, liver and adipose tissue.

An increased glucose uptake (i.e., higher glucose tolerance) may result from various processes, such as higher amounts of insulin secreted from the pancreatic beta-cells after a meal, an increased insulin sensitivity of peripheral tissues, or an increase in (non)-insulin-mediated glucose uptake via the recruitment of glucose transporters in exercising muscles.

Glucose tolerance: The relative amount of glucose taken up by peripheral tissues.

Insulin resistance: The diminished ability of cells to respond to the action of insulin in transporting glucose (sugar) from the bloodstream into muscle and other tissues.

Basal Glucose Levels

Blood glucose concentrations peak right before the activity period. Humans show high glucose output and insulin requirements in the early morning hours, suggesting an anticipatory increase of glucose metabolism for the upcoming activity period [4]. Also in rats a daily variation in glucose metabolism was observed. Again, the highest glucose tolerance and insulin sensitivity was observed at the time of awakening. Diminished insulin sensitivity and decreased insulin responses to a glucose load have both been suggested to be responsible for the reduced glucose tolerance later in the day [5–7]. However, it is still unclear which tissues are responsible for these changes in glucose tolerance.

Numerous studies have shown that the suprachiasmatic nucleus (SCN) of the hypothalamus influences behavioral rhythms such as the sleep–wake cycle, locomotor activity and feeding behavior [8]. There is also clear evidence for the involvement of the SCN in anterior pituitary hormonal release [8], and lately it has become clear that the SCN also deploys the autonomic nervous system (ANS) to spread its message to the peripheral organs (liver, pancreas, fat, adrenal, etc.) [9–15].

The first evidence that the SCN is involved in the daily rhythm in glucose metabolism came from the work of Nagai and Nakagawa: (1) SCN lesions abolish the daily rhythms in plasma concentrations of glucose and insulin [16] and (2) a pronounced day–night difference exists in the response to 2-deoxy glucose, a glucose-utilization inhibitor [17]. SCN-lesioned rats, however, do not have a rhythm in food intake [18], and an indirect effect on glucose metabolism through the lack of a feeding rhythm could thus not be excluded, as food intake increases glucose concentrations. Indeed, in rodents, restricted daytime feeding will prompt the glucose and insulin peaks to shift to the daytime [19], and there is therefore every possibility that the SCN-driven rhythm in food intake causes a daily rhythm in plasma concentrations of glucose and insulin. To test whether there is a direct influence of the SCN on glucose metabolism which is independent from its effect on feeding behavior, a series of experiments using a 6-meals-a-day feeding schedule (to avoid interference from the day–night rhythm in feeding activity) was performed.

The first experiments with this scheduled feeding regimen in rats revealed a clear diurnal rhythm in meal-induced responses in glucose and insulin [20] (Fig. 7.1). The data on the pre-meal glucose values from this study suggested a daily rhythm in basal plasma glucose concentrations independent of the feeding pattern as well. When food availability is limited to a fixed time of the day, it causes a phase advance in circadian rhythms, such as locomotor activity, and it changes the daily rhythm in plasma corticosterone concentrations [21–23]. Thus, because food itself serves as a strong synchronizer/entraining signal, we measured locomotor activity patterns as well as plasma corticosterone concentrations (which affect glucose metabolism by affecting insulin sensitivity or glucose output from the liver) in rats on a 6-meals-a-day feeding schedule. Locomotor activity patterns and plasma corticosterone concentrations did not change in rats on the 6-meal scheduled feeding regimen, suggesting that this regimen disrupts the daily pattern in feeding activity but not overall behavior [20, 24]. That the SCN directly mediates the daily rhythm in plasma glucose concentrations became evident when subsequent experiments were performed with SCN-lesioned and SCN-intact rats on this scheduled feeding regimen [24] (Fig. 7.1). The daily rhythm observed in plasma glucose concentrations nicely resembled the observed daily rhythm in animals fed ad libitum, suggesting a direct influence of the SCN on plasma glucose concentrations, independent of the feeding rhythm. The maintained rhythmicity in plasma concentrations during fasting is also in agreement with a direct influence of the SCN on glucose metabolism independent of its effect on feeding behavior. Although there are reports of a circadian rhythm in insulin concentrations [25], we did not find an obvious and direct influence of the SCN on basal plasma insulin concentrations, independent of feeding activity: the plasma insulin concentrations increased after every meal. However, when considering the pre-meal values, only 50% of the SCN-intact rats on the scheduled feeding regimen showed a 24-h rhythm of plasma insulin concentrations. Both rhythms

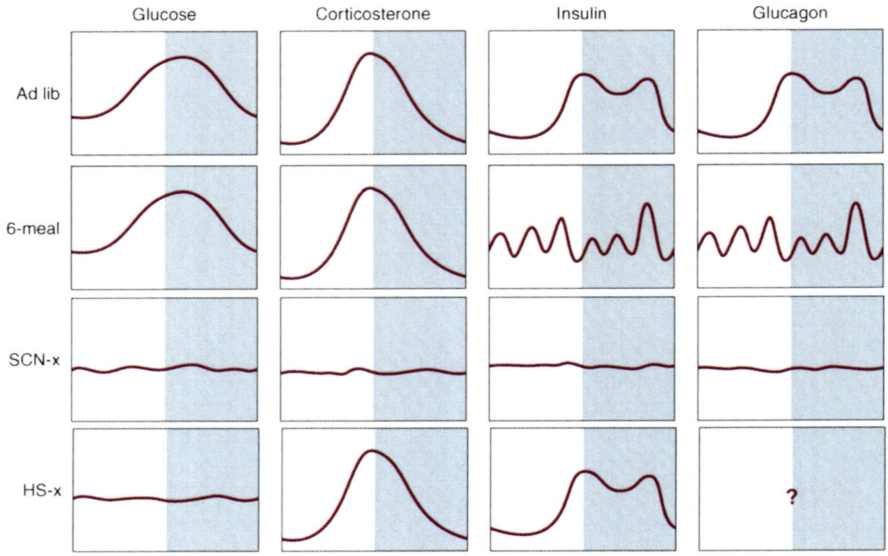

Fig. 7.1 In ad libitum feeding animals (Ad lib) plasma glucose concentrations rise prior to the activity period (*gray area*). Because plasma glucose concentrations are influenced by the rhythm in food intake, we measured plasma glucose concentrations in animals on a 6-meals-a-day schedule (6-meal) who still had a clear day–night rhythm in plasma glucose concentrations. The day–night rhythm in plasma glucose concentrations is dependent on the SCN, because SCN-lesioned animals (SCN-x) on a 6-meals-a-day schedule no longer exhibit a day–night rhythm in plasma glucose concentrations. Moreover, the day–night rhythm is also dependent on the hepatic sympathetic innervation, because animals with a sympathetic liver denervation (HSx) do not exhibit a rhythm either. The hormones important for glucose metabolism—corticosterone, insulin, and glucagon—do change under different conditions, but these changes do not explain the day–night rhythm in plasma glucose concentrations (adapted from [112])

were abolished with SCN lesions [24]. Moreover, fasted rats do not show such a daily rhythm in plasma insulin concentrations [24, 26], which supports the hypothesis that feeding behavior and/or glucose concentrations are more important than a direct SCN influence on basal insulin secretion. This enabled us to ascertain that the SCN increases plasma glucose concentrations before waking (Box 7.1).

The increased plasma glucose concentrations could be derived from a decrease in glucose uptake or from an increase in glucose output, Interestingly, however, a series of timed intravenous glucose tolerance tests showed glucose uptake to have a clear 24-h rhythm with the highest uptake at the end of the light period as well [27]. This rhythm, too, depends on the SCN and not on the rhythm in food intake. It is obvious that the simultaneous rise in plasma glucose concentrations and glucose uptake can only occur when glucose output to the general circulation is such that it compensates for the increased glucose uptake. A similar situation occurs in humans; before awakening, glucose production and glucose concentrations are increased while, at the same time, glucose utilization is high [28]. Consequently, the increase in plasma glucose concentrations before the onset of activity is due to increased glucose production, not to a decreased glucose utilization.

The main source for glucose is the liver. Interestingly, anatomical tracing experiments revealed that there are connections between the SCN and the liver, via the autonomic nervous system (ANS) [14, 29]. These connections may well be involved in the organization of the 24-h rhythm in plasma glucose concentrations. The involvement of the ANS in SCN-driven changes in glucose metabolism was first suggested by Nagai and colleagues. They showed that electrical stimulation of the SCN resulted in hyperglycemia. This effect was prevented by blocking the output of the autonomic nervous system with (i.p.) administration of α and β-adrenergic blockers [30, 31]. Administering an adrenergic blocker i.p., however, is not specific and more recent experiments therefore used hepatic sympathetic denervations to determine the role of autonomic innervation of the liver for the generation of the daily rhythm in plasma glucose concentrations. From these studies it became clear that the SCN indeed requires an intact sympathetic innervation of the liver to generate a daily rhythm in plasma glucose concentrations [32, 33].

The SCN does not innervate the autonomic motor neurons in the brainstem or spinal cord directly, but transmits its signal to other areas within the hypothalamus. For the SCN to transmit its rhythms, the paraventricular nucleus of the hypothalamus (PVN) is the most important area to affect autonomic signaling to peripheral organs. The PVN receives signals from the SCN and has extensive projections to sympathetic and parasympathetic motor neurons in the spinal cord and in the brainstem, respectively [34–37]. The functional importance of this SCN–PVN connection in controlling plasma glucose concentrations was revealed by administering different SCN transmitter agonists and antagonists into the vicinity of the PVN [29]. The most pronounced effects on plasma glucose concentrations were observed after the administration of either bicuculline (BIC; a GABA-A antagonist) or NMDA (an agonist of glutamatergic receptors) to the PVN: both resulted in a prolonged and significant increase in plasma glucose concentrations. Both drugs also increased plasma glucagon concentrations (which may stimulate glucose output) but did not affect plasma insulin concentrations in any significant way. Blockade of GABA-ergic receptors resulted in increased plasma concentrations of corticosterone, which—like glucagon—is known to increase gluconeogenesis, whereas stimulating the glutamate receptors did not. These data indicate that it is unlikely that the hyperglycemia induced by the stimulation of PVN neurons is a result of changes in either insulin or corticosterone release, although increased glucagon release could be a causative factor. Later experiments showed that prior selective denervation of the sympathetic, but not the parasympathetic, autonomic input to the liver completely prevented the hyperglycemic effects of both BIC and NMDA [29, 38]. The hyperglycemic effects disappeared in the sympathetic denervated animals, notwithstanding pronounced increases of plasma concentrations of glucagon and corticosterone. Together, these functional studies demonstrate that stimulating neuronal activity in the PVN results in hyperglycemia through activating sympathetic input to the liver. Repeating the above experiments at different times of the day and in SCN-lesioned animals confirmed the SCN as the major site of origin for the GABA and glutamatergic inputs to the PVN [39].

The experiments described above show clear effects on plasma glucose concentrations but fail to provide evidence for an effect of the hypothalamus on glucose production via the autonomic innervation of the liver. A more recent study showed that local administration of BIC in the hypothalamus increases hepatic glucose production when injected in the PVN or the dorsomedial hypothalamus (DMH), which both receive SCN input. The most effective area for increasing hepatic glucose production, however, was the perifornical area (PF) [40], another area receiving input from the SCN [41]. Thus, in order to be able to generate the 24 h rhythm in plasma glucose concentrations, the SCN transmits its signal to a number of hypothalamic areas important for the regulation of glucose metabolism.

Acute Changes in Glucose Levels

In addition to the SCN-generated day–night rhythm in plasma glucose concentrations, the brain also responds to acute changes in plasma glucose concentrations by increasing glucose production when blood glucose levels drop, and by increasing glucose uptake and inhibiting glucose production when glucose levels rise. Circulating insulin levels, which increase when glucose concentrations increase, may also act on the brain and influence glucose metabolism. Glucose-sensing neurons, which are either activated or inhibited by changes in extracellular glucose concentrations, are located throughout the hypothalamus. For example, in the arcuate nucleus (ARC), neurons expressing neuropeptide Y (NPY) have been shown to be glucose-inhibitory neurons. But these NPY neurons also respond to insulin, which, like glucose, also has an inhibitory effect on the neural activity of NPY neurons [42]. Since the discovery of leptin in 1994 [43], the ARC has attracted a lot of attention. It is in close contact to the circulation because of a "leaky" blood brain barrier, and has extensive projections to several areas within the hypothalamus. For example, projections to the lateral hypothalamus (LH) and the ventromedial hypothalamus (VMH) are part of this neurocircuitry in the hypothalamus that regulates glucose metabolism. Originally the LH was described as a hunger center, because cats with lesions in this area lost their appetite and lost weight dramatically. Furthermore, it is an area that contains many glucose-inhibitory neurons. On the other hand, when the VMH, was lesioned in cats, they did not become satiated resulting in massive obesity, as well as insulin-resistance. In addition, they had more glucose excitatory neurons than glucose inhibitory neurons. The VMH is also proposed to play an important role in the response to hypoglycemia, the so-called counterregulatory response (for a review on hypothalamic glucose sensing neurons: [44, 45]). Several neuropeptides within the hypothalamic areas just mentioned have been shown to be involved in the regulation of glucose production and glucose uptake, which are summarized in Fig. 7.2.

With regard to the control of rhythms in glucose metabolism by hypothalamic neuropeptides, the orexins deserve special attention. Orexin-A and orexin-B (also known as hypocretin 1 and 2) were initially identified as endogenous ligands for an

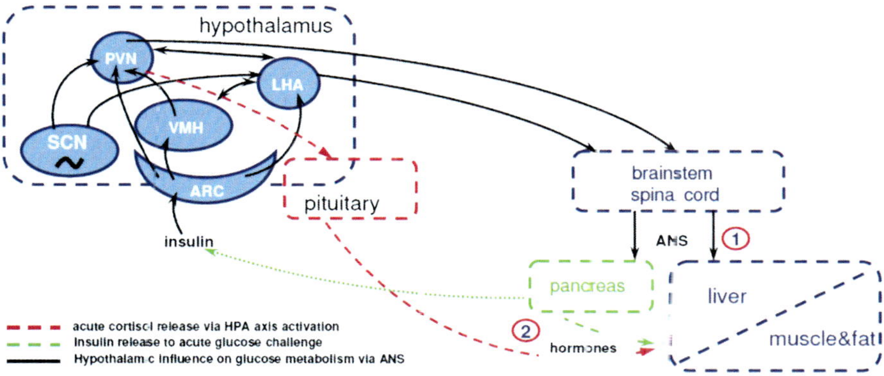

Fig. 7.2 A schematic view of the brain's network that maintains plasma glucose concentrations between strict boundaries. (*1*) The hypothalamic areas, where numerous neuropeptides are localized, influence glucose metabolism via the autonomic nervous system (*solid black lines*). (2) Glucose metabolism is also influenced by hormones: (a) via the pituitary route (*dotted red line*), which with acute challenges, such as stress, will release corticosterone to influence hepatic glucose output and (b) via the pancreas that will sense glucose concentrations resulting in insulin release or glucagon release to regulate glucose levels. However, the pancreas also receives signals from the brain via the autonomic nervous system (*dotted green lines*). In turn, insulin also acts on its receptors in the ARC to regulate glucose metabolism. *Abbreviations*: *ANS* autonomic nervous system, *ARC* arcuate nucleus, *HPA* hypothalamus–pituitary–adrenal, *LHA* lateral hypothalamic area, *PVN* (hypothalamic) paraventricular nucleus, *SCN* suprachiasmatic nucleus, *VMH* ventromedial nucleus of the hypothalamus

orphan receptor involved in narcolepsy [46, 47], but were later shown to be important for the maintenance of wakefulness. Orexin neurons were shown to be exclusively localized in the lateral hypothalamus and, in addition to their importance for maintaining wakefulness, also turned out to be involved in the regulation of feeding behavior and energy metabolism. More recent data show that an increased availability of orexin in the central nervous system, either by i.c.v. infusion or by local administration or activation of orexin neurons (using GABA antagonist) increases plasma glucose concentrations through an increase in hepatic glucose production [40]. Moreover, orexin is also able to stimulate glucose uptake in skeletal muscle via its action in the VMH [48].

The stimulatory effect of orexin on EGP could be blocked by a hepatic sympathetic denervation but not by a parasympathetic one [40]. Moreover, other studies clearly showed that the activity of the orexin neurons is under tight control of a GABAergic input that is probably derived from the circadian system [29]. These data indicate that the circadian rhythm in orexin release [49] might be implicated in the genesis of the daily rhythm in plasma glucose concentrations as described above. Indeed, i.c.v. (but not i.v.) administration of an orexin antagonist during the final 8 h of the light period completely blocked the endogenous increase in glucose appearance (probably glucose production) until the start of the dark period. Once the animals started eating in the dark period, glucose appearance also increased in the i.c.v. orexin-antagonist treated animals. These findings strongly suggest that the orexin neurons in the LH are an important link in the circadian control of the SCN, via the autonomic nervous system, over the daily peripheral glucose rhythm.

CNS Control of Hormones Affecting Glucose Metabolism

In addition to its role in activating the sympathetic input to the liver, hepatic glucose production is also stimulated by corticosterone and glucagon, whereas increased circulating plasma insulin levels will inhibit hepatic glucose production (Box 7.1). Like plasma glucose concentrations, also the magnitude of an endocrine response to a glucose challenge varies over the activity/inactivity cycle. For example, a marked effect of time of day on neuroendocrine responses to prolonged moderate exercise was reported in healthy volunteers [50] and an oral glucose load in the early morning hours gives a higher insulin response compared to glucose given in the evening or afternoon [6]. The same was found in rats with meals equally distributed over the light/dark-cycle (i.e., insulin responses varied depending on the time of the day the meal was consumed, despite equal meal sizes) [20]. In addition, SCN-lesion studies showed this variation in endocrine function to be dependent on the SCN [18]. Furthermore, in rats, the corticosterone response to a metabolic stressor, such as an insulin injection, is higher at the end of the light period than at the beginning of the light period [51]. 2-Deoxy-glucose (2-DG), a glucose inhibitor, which stimulates glucose output, varies its corticosterone responses with the time of day [114]. The central nervous system is known to influence endocrine responses via (1) neuroendocrine neurons leading to anterior pituitary hormonal release (corticosterone, growth hormone) and (2) neurons of the ANS innervating peripheral organs such as the adrenal and pancreas (adrenalin, insulin, and glucagon). For the daily variation in the response to an acute glucose challenge, the SCN uses both routes to determine the magnitude of the endocrine response. There is evidence for the involvement of the SCN in anterior pituitary hormonal release [8], and neural connections are in place between the SCN and endocrine organs such as the pancreas and adrenal [10, 11, 13]. It has been shown that the SCN influences corticosterone secretion via two routes: the SCN affects CRH release then stimulating ACTH release from the pituitary and subsequently the secretion of corticosterone from the adrenal cortex and it is able to set the sensitivity of the adrenal gland to ACTH via the sympathetic innervation of the adrenal [10, 52–54]. Because corticosterone increases gluconeogenesis, it may well be that, in general, the SCN sets the endocrine system to respond to glucose challenges via both the anterior pituitary and the ANS. Furthermore, the daily variation in meal-induced insulin responses is eliminated by atropine, which blocks the parasympathetic system [18], suggesting that the SCN influences insulin responses via neural projections to the pancreas. Thus, the SCN influences hormone secretion to an acute glucose challenge via an interplay between hormonal pituitary release and modulation of the ANS input to the organs involved, such as pancreas and adrenal. A hormonal involvement in the SCN-regulated anticipatory increase in plasma glucose concentrations towards the beginning of the activity period, however, is unlikely. Insulin concentrations rise after every meal in a scheduled feeding regimen, and decline during fasting, i.e. contrary to the plasma glucose concentrations that remain rhythmic in these conditions [24, 26]. In addition, plasma glucagon concentrations increase after every meal in these rats when subjected to the scheduled feeding regimen and do not show a daily pattern that could explain the rhythm in plasma glucose concentrations [55] (Fig. 7.1). The peak in plasma glucose concentrations occurs at the same time as the circadian peak in plasma corticosterone concentrations [56–60]. Because corticosterone increases glucose output, it has been suggested to be responsible for the rise in glucose concentrations at the end of the light period. However, in humans it was shown that blocking the synthesis of glucocorticoids does not prevent the rise in glucose at the end of the inactivity period [61]. In addition, a selective denervation of the sympathetic innervation of the liver eliminates the daily rhythm in plasma glucose concentrations, despite an intact daily rhythm of plasma corticosterone [62] (Fig. 7.1). Two other hormones that respond to a decrease in plasma glucose concentrations and may be important for the daily rhythm in glucose metabolism are growth hormone and adrenalin. However, in rats neither of these hormones adheres to a 24-h rhythm that is likely to account for the 24-h rhythm in plasma glucose concentrations. The 24-h rhythm in growth hormone in rats is completely different from that in humans (i.e. male rats do not display a clear circadian rhythm in growth hormone concentrations, but instead show a strong ultradian rhythm with a 3-h period [63, 64]). Although adrenalin rapidly increases hepatic glucose production, its daily rhythm does not correlate with the rhythm in glucose levels, but rather correlates more strongly with locomotor activity levels [57]. Thus, in situations of acute glucose challenge, the SCN sets the response of the endocrine system through its effects on the neuroendocrine and autonomic nervous system, whereas the anticipatory increase in plasma glucose concentrations in the early morning seems to be completely dependent on the SCN control via the autonomic nervous system.

In conclusion, basal plasma glucose concentrations fluctuate with the light/dark-cycle in a rhythm generated by the hypothalamic biological clock in the SCN, which, through its influence on the perifornical orexin system and the autonomic nervous system transmits its signal to the liver to stimulate glucose production, resulting in a rise in plasma glucose prior to the activity period. At the same time the SCN also increases glucose tolerance via an as yet unknown pathway, although probably, again, the perifornical orexin system and the autonomic nervous system are involved. On the other hand, acute changes in plasma glucose concentrations are perceived by glucose sensing neurons in different hypothalamic areas, or reported indirectly via changes in insulin concentrations, and subsequently glucose concentrations are returned to basal via activation of glucose production by autonomic nerves innervating the liver, or by hormones released from glands such as the pancreas and/or adrenals.

Insulin Resistance and Type 2 Diabetes Mellitus

Both insulin resistance and β-cell failure contribute to the pathogenesis and progression of T2DM as they both precede the development of hyperglycemia [65]. Insulin resistance in T2DM primarily manifests itself at the level of the liver and the skeletal muscle, resulting in impaired insulin-stimulated glucose uptake and a failure to adequately suppress hepatic glucose production [66]. On the other hand, β-cell failure (defective function and loss of cell mass) results in altered glucose-stimulated insulin secretion [67, 68]. The disruption of insulin action and secretion are influenced by both environmental and genetic factors of which environmental factors, such as obesity and physical activity, are more linked to insulin action [69] and the genetics are more linked to β-cell function [70, 71].

Environmental Factors

Epidemiological studies show an association between disturbances in the sleep–wake rhythm and the occurrence of T2DM. For example, shift workers run a higher risk of developing obesity and T2DM [72–74] and it has been proposed that this is due to a chronic misalignment between the endogenous central clock and the behavioral sleep–wake and fasting–feeding rhythms. That circadian misalignment may indeed result in metabolic disturbances was strongly supported by a recent study performed in a laboratory setting in which subjects were subjected to an 11-day protocol consisting of repeated 28-h "days," with four standardized meals per 28-h day. The misalignment between the endogenous ~24-h day and the enforced behavioral 28-h day resulted in increased plasma concentrations of glucose and insulin, increased blood pressure and decreased plasma leptin concentrations. Moreover, in three out of the ten subjects this circadian disorganization pushed the postprandial glucose response into the pre-diabetic range [75]. Since this severe misalignment, mimicking shift work, only occurs in a small portion of the population it probably does not account for the increase in prevalence of T2DM in Western Society. However, it could well be that a milder form of circadian misalignment contributes to the increased rates of T2DM. With a simple questionnaire Roenneberg et al. studied the sleep–wake times of thousands of people all over the world and determined their chronotype [76, 77], the upshot of which was that more than half of the population lives with a biological clock that is permanently out of phase with the environmental time, a condition they refer to as "social jetlag." Interestingly, the magnitude of the social jetlag correlates with an increased prevalence of the metabolic syndrome. Like the small and the more severe misalignments between the endogenous clock and the sleep–wake cycle, sleep deprivation and poor sleep, too, have been shown to reduce insulin sensitivity (for further reading on this topic: [78]).

In addition to the above described changes in sleep–wake cycles, changes in the feeding rhythm may also influence glucose metabolism. Clock genes, the essential elements of the molecular clocks in the neurons of the central clock, are also present in peripheral tissues, including those tissues that are important for glucose metabolism, such as the liver, pancreas, and adipocytes [79–81]. The circadian system is thus regulated by the master clock located in the SCN, which orchestrates the circadian oscillations within the organs to drive tissue specific outputs and synchronize them with the light dark cycle [82]. Feeding patterns are interesting in this respect, because it has been shown that a restricted, defined feeding time during the light period in nocturnal rodents such as rats and mice shifts the clock gene rhythms in the liver, but not in the central clock in the SCN, thus causing misalignment between the central clock and the peripheral clocks [83]. Since clock genes in the liver are tightly linked to genes involved in glucose metabolism it is very likely that this misalignment will have consequences for glucose metabolism [84–86].

It has also been postulated that a high-fat diet in mice alters the function of the circadian clock, with an altered period of the locomotor activity rhythm and changes in the rhythms of the core clock genes in adipose and liver tissue [87–90]. It has therefore been hypothesized that, via alterations in circadian rhythmicity, a high-fat diet contributes to the development of obesity and insulin resistance. That nutritional status may affect obesity and glucose metabolism via altered circadian physiology was also indicated by a recent study in which protein-restricted male C57BL/6J offspring developed increased adiposity and glucose intolerance and exhibited altered circadian physiology and circadian clock gene expression profiles before the onset of obesity [91]. However, whether this is specific to a fetal nutritional status remains to be investigated. Ando et al. [92] demonstrated that rhythmic mRNA expression of clock genes is not dampened in the visceral adipose tissue of non-obese diabetic GK rats, and that impairment of molecular clock function in the visceral adipose tissue may thus be related to obesity, but not directly to diabetes. This was also supported by a study conducted in humans, in which several fat biopsies were collected over the day–night cycle and no changes in rhythms of clock genes were observed in white adipose tissue between diabetic and nondiabetic subjects [93]. Therefore, although high fat feeding and/or obesity may alter circadian physiology, it is unclear whether this circadian change underlies the development of insulin resistance associated with obesity.

Clock Genes in Glucose Metabolism

Disruption of core clock genes alters energy metabolism in many ways. This is not surprising, since the molecular clock genes (Clock, Bmal1, Cry 1–2, Per 1–3, and Rev-erbα) control the expression and activity of numerous enzymes, transport systems and nuclear receptors involved in energy metabolism, in organs such as liver, pancreas, and white adipose tissue [113].

Turek et al. [94] showed that homozygous in cursive mutant C57BL/6J mice are obese with several characteristics of metabolic syndrome including hepatic steatosis, hyperglycemia, hypertriglyceridemia, hypercholesterolemia, and hyperleptinemia.

Interestingly, however, these mice were not hyperinsulinemic as would be expected with hepatic steatosis and hyperglycemia, but they were more insulin-sensitive at a young age [95]. In cursive disruption in ob/ob mice resulted in more body weight gain, higher plasma triglycerides, and more cholesterol accumulation compared to ob/ob mice without this disruption [96]. In cursive mutants, as well as in cursive mutants, also show impaired glucose tolerance, reduced insulin secretion and defects in size and proliferation of pancreatic islets that worsen with age. In addition, it was shown that Clock disruption alters expression of islet genes involved in growth, survival and synaptic vesicle assembly. Two research groups went on to target Bmal1 specifically in the beta cell and indeed showed that conditional ablation of the pancreatic clock causes diabetes mellitus due to defective β-cell function [95, 97]. Very recently, it was shown that Rev-erb alpha down-regulation (using RNA interference) in islet cells and MIN-6 cells impaired glucose-induced insulin secretion [98]. The importance of pancreatic clock genes for the development of type 2 diabetes is further supported by the variant recently found near CRY2, associated with elevated fasting glucose in humans [99, 100].

It should be noted that the obesity observed in clock mutants much depends on their genetic background. For example, on an ICR background, in cursive mutation does not result in obesity, but these mice do have a defective fat absorption system [101]. Interestingly, in the study by Kennaway et al. [102], it was shown that in cursive + MEL mutant mice, crossed with a CBA strain, were not obese either, but were insulin sensitive and exhibited normal fasting glucose levels. The CBA strain has the same origin as C3H mice that are considered "normal," i.e., without changes in enzymes important for glucose metabolism, which has been shown for the C57BL/6J background [103]. Moreover, the majority of null mutations in the circadian clock components are associated with leanness, despite normophagia or hyperphagia [104–106], suggesting that global clock gene disruption affects energy metabolism in a complex manner.

The liver plays an important role in the circadian rhythm in glucose metabolism. It has been shown that restricted feeding shifts rhythmic patterns of clock genes in the liver and thus uncouples it from the central clock—where no changes in rhythmic patterns of clock genes were observed. Over 350 circadian transcripts were identified in the liver, of which 10%, including the core gene Per2, maintain rhythmicity in the absence of a functional hepatocyte clock, supporting a role for behavioral rhythms (eating, sleeping, locomotor activity) in liver gene rhythms [107]. More precise roles for the core clock genes in hepatic glucose metabolism have been clearly shown using tissue specific clock knock outs; this has been nicely reviewed by Bray and Young [108]. In short, liver-specific Bmal1 disruption in mice increases glucose tolerance, with normal insulin production and normal body fat content [109]. Rev-erb α directly regulates the expression of multiple gluconeogenic enzymes in liver, including glucose-6-phosphatase and phosphoenylpyruvate [110]. And liver-specific Cry KO inhibits glucagon-induced gluconeogenesis, whereas overexpression of Cry protein in the liver of diabetic db/db mice improves glucose tolerance [111].

The evidence until now thus supports a role for pancreatic clock genes in the development of T2DM, and for liver clock genes in the control of glucose tolerance through their effect on hepatic gluconeogenesis.

Overall Conclusion

The field of chronobiology is making great strides. Exciting progress has been made in the understanding of how central and peripheral circadian clocks influence glucose metabolism. We discussed the role of the central clock in the regulation of the day–night rhythms in blood glucose and insulin sensitivity. In addition, we reviewed the role of clock genes in peripheral tissues that are essential for insulin secretion and gluconeogenesis. It is clear that disturbances in central or peripheral clock functioning will result in altered glucose metabolism and that these alterations may lead to the development of T2DM. However, much is still to be investigated regarding the environmental factors that may alter circadian physiology

Summary Points

- Blood glucose concentrations display a clear daily rhythm enforced by the biological clock and the autonomic nervous system.
- The acute hormonal response to a glucose challenge depends on the time of the day.
- Clock gene disruption results in hypoinsulinemia, but its effects on insulin sensitivity are less clear and might depend on the response of body weight regulation.
- High-fat feeding and obesity alter circadian physiology, but it is unclear whether this contributes to the development of diabetes.

References

1. Feldberg W, Pyke D, Stubbs WA (1985) Hyperglycaemia: imitating Claude Bernard's piqure with drugs. J Auton Nerv Syst 14(3):213–228
2. Kreier F, Swaab DF (2010) Chapter 23: history of neuroendocrinology "the spring of primitive existence". Handb Clin Neurol 95:335–360
3. Cannon W (1929) Organization for physiological homeostasis. Physiol Rev 9(3):399–431
4. Cuesta M, Clesse D, Pevet P, Challet E (2009) From daily behavior to hormonal and neurotransmitters rhythms: comparison between diurnal and nocturnal rat species. Horm Behav 55(2):338–347
5. Carroll KF, Nestel PJ (1973) Diurnal variation in glucose tolerance and in insulin secretion in man. Diabetes 22(5):333–348
6. Lee A, Ader M, Bray GA, Bergman RN (1992) Diurnal variation in glucose tolerance. Cyclic suppression of insulin action and insulin secretion in normal-weight, but not obese, subjects. Diabetes 41(6):750–759

7. Whichelow MJ, Sturge RA, Keen H, Jarrett RJ, Stimmler L, Grainger S (1974) Diurnal variation in response to intravenous glucose. Br Med J 1(5906):488–491
8. Rusak B, Zucker I (1979) Neural regulation of circadian rhythms. Physiol Rev 59(3):449–526
9. Bartness TJ, Bamshad M (1998) Innervation of mammalian white adipose tissue: implications for the regulation of total body fat. Am J Physiol 275(5 Pt 2):R1399–R1411
10. Buijs RM, Wortel J, Van Heerikhuize JJ, Feenstra MG, Ter Horst GJ, Romijn HJ, Kalsbeek A (1999) Anatomical and functional demonstration of a multisynaptic suprachiasmatic nucleus adrenal (cortex) pathway. Eur J Neurosci 11(5):1535–1544
11. Buijs RM, Chun SJ, Niijima A, Romijn HJ, Nagai K (2001) Parasympathetic and sympathetic control of the pancreas: a role for the suprachiasmatic nucleus and other hypothalamic centers that are involved in the regulation of food intake. J Comp Neurol 431(4):405–423
12. Buijs RM, la Fleur SE, Wortel J, van Heijningen C, Zuiddam L, Mettenleiter TC, Kalsbeek A, Nagai K, Niijima A (2003) The suprachiasmatic nucleus balances sympathetic and parasympathetic output to peripheral organs through separate preautonomic neurons. J Comp Neurol 464(1):36–48
13. Jansen AS, Hoffman JL, Loewy AD (1997) CNS sites involved in sympathetic and parasympathetic control of the pancreas: a viral tracing study. Brain Res 766(1–2):29–38
14. la Fleur SE, Kalsbeek A, Wortel J, Buijs RM (2000) Polysynaptic neural pathways between the hypothalamus, including the suprachiasmatic nucleus, and the liver. Brain Res 871(1):50–56
15. Ueyama T, Krout KE, Nguyen XV, Karpitskiy V, Kollert A, Mettenleiter TC, Loewy AD (1999) Suprachiasmatic nucleus: a central autonomic clock. Nat Neurosci 2(12):1051–1053
16. Yamamoto H, Nagai K, Nakagawa H (1987) Role of SCN in daily rhythms of plasma glucose, FFA, insulin and glucagon. Chronobiol Int 4(4):483–491
17. Yamamoto H, Nagai K, Nakagawa H (1984) Bilateral lesions of the SCN abolish lipolytic and hyperphagic responses to 2DG. Physiol Behav 32(6):1017–1020
18. Strubbe JH, Prins AJ, Bruggink J, Steffens AB (1987) Daily variation of food-induced changes in blood glucose and insulin in the rat and the control by the suprachiasmatic nucleus and the vagus nerve. J Auton Nerv Syst 20(2):113–119
19. Escobar C, Diaz-Munoz M, Encinas F, Aguilar-Roblero R (1998) Persistence of metabolic rhythmicity during fasting and its entrainment by restricted feeding schedules in rats. Am J Physiol 274(5 Pt 2):R1309–R1316
20. Kalsbeek A, Strubbe JH (1998) Circadian control of insulin secretion is independent of the temporal distribution of feeding. Physiol Behav 63(4):553–558
21. Challet E, Pevet P, Vivien-Roels B, Malan A (1997) Phase-advanced daily rhythms of melatonin, body temperature, and locomotor activity in food-restricted rats fed during daytime. J Biol Rhythms 12(1):65–79
22. Choi S, Wong LS, Yamat C, Dallman MF (1998) Hypothalamic ventromedial nuclei amplify circadian rhythms: do they contain a food-entrained endogenous oscillator? J Neurosci 18(10):3843–3852
23. Kalsbeek A, Van Heerikhuize JJ, Wortel J, Buijs RM (1998) Restricted daytime feeding modifies suprachiasmatic nucleus vasopressin release in rats. J Biol Rhythms 13(1):18–29
24. la Fleur SE, Kalsbeek A, Wortel J, Buijs RM (1999) A suprachiasmatic nucleus generated rhythm in basal glucose concentrations. J Neuroendocrinol 11(8):643–652
25. Boden G, Ruiz J, Urbain JL, Chen X (1996) Evidence for a circadian rhythm of insulin secretion. Am J Physiol 271(2 Pt 1):E246–E252
26. Dallman MF, Akana SF, Bhatnagar S, Bell ME, Choi S, Chu A, Horsley C, Levin N, Meijer O, Soriano LR et al (1999) Starvation: early signals, sensors, and sequelae. Endocrinology 140(9):4015–4023
27. la Fleur SE, Kalsbeek A, Wortel J, Fekkes ML, Buijs RM (2001) A daily rhythm in glucose tolerance: a role for the suprachiasmatic nucleus. Diabetes 50(6):1237–1243
28. Bolli GB, Gerich JE (1984) The "dawn phenomenon"—a common occurrence in both non-insulin-dependent and insulin-dependent diabetes mellitus. N Engl J Med 310(12):746–750

29. Kalsbeek A, la Fleur SE, van Heijningen C, Buijs RM (2004) Suprachiasmatic GABAergic inputs to the paraventricular nucleus control plasma glucose concentrations in the rat via sympathetic innervation of the liver. J Neurosci 24(35):7604–7613
30. Fujii T, Inoue S, Nagai K, Nakagawa H (1989) Involvement of adrenergic mechanism in hyperglycemia due to SCN stimulation. Horm Metab Res 21(12):643–645
31. Nagai K, Fujii T, Inoue S, Takamura Y, Nakagawa H (1988) Electrical stimulation of the suprachiasmatic nucleus of the hypothalamus causes hyperglycemia. Horm Metab Res 20(1):37–39
32. Cailotto C, van Heijningen C, van der Vliet J, van der Plasse G, Habold C, Kalsbeek A, Pevet P, Buijs RM (2008) Daily rhythms in metabolic liver enzymes and plasma glucose require a balance in the autonomic output to the liver. Endocrinology 149(4):1914–1925
33. la Fleur SE (2001) The suprachiasmatic nucleus generated rhythm in blood glucose. Thesis, University of Amsterdam, the Netherlands
34. Buijs RM (1978) Intra- and extrahypothalamic vasopressin and oxytocin pathways in the rat. Pathways to the limbic system, medulla oblongata and spinal cord. Cell Tissue Res 192(3):423–435
35. Jansen AS, Farwell DG, Loewy AD (1993) Specificity of pseudorabies virus as a retrograde marker of sympathetic preganglionic neurons: implications for transneuronal labeling studies. Brain Res 617(1):103–112
36. Luiten PG, Ter Horst GJ, Karst H, Steffens AB (1985) The course of paraventricular hypothalamic efferents to autonomic structures in medulla and spinal cord. Brain Res 329(1–2):374–378
37. Swanson LW, McKellar S (1979) The distribution of oxytocin- and neurophysin-stained fibers in the spinal cord of the rat and monkey. J Comp Neurol 183(1):87–106
38. Kalsbeek A, Ruiter M, la Fleur SE, Cailotto C, Kreier F, Buijs RM (2006) The hypothalamic clock and its control of glucose homeostasis. Prog Brain Res 153:283–307
39. Kalsbeek A, Foppen E, Schalij I, van Heijningen C, van de Vliet J, Fliers E, Buijs RM (2008) Circadian control of the daily plasma glucose rhythm: an interplay of GABA and glutamate. PLoS One 3(9):e3194
40. Yi CX, Serlie MJ, Ackermans MT, Foppen E, Buijs RM, Sauerwein HP, Fliers E, Kalsbeek A (2009) A major role for perifornical orexin neurons in the control of glucose metabolism in rats. Diabetes 58(9):1998–2005
41. Vrang N, Mikkelsen JD, Larsen PJ (1997) Direct link from the suprachiasmatic nucleus to hypothalamic neurons projecting to the spinal cord: a combined tracing study using cholera toxin subunit B and Phaseolus vulgaris-leucoagglutinin. Brain Res Bull 44(6):671–680
42. van den Top M, Lee K, Whyment AD, Blanks AM, Spanswick D (2004) Orexigen-sensitive NPY/AgRP pacemaker neurons in the hypothalamic arcuate nucleus. Nat Neurosci 7(5):493–494
43. Zhang Y, Proenca R, Maffei M, Barone M, Leopold L, Friedman JM (1994) Positional cloning of the mouse obese gene and its human homologue. Nature 372(6505):425–432
44. Burdakov D, Gonzalez JA (2009) Physiological functions of glucose-inhibited neurones. Acta Physiol (Oxf) 195(1):71–78
45. Levin BE, Routh VH, Kang L, Sanders NM, Dunn-Meynell AA (2004) Neuronal glucosensing: what do we know after 50 years? Diabetes 53(10):2521–2528
46. de Lecea L, Kilduff TS, Peyron C, Gao X, Foye PE, Danielson PE, Fukuhara C, Battenberg EL, Gautvik VT, Bartlett FS et al (1998) The hypocretins: hypothalamus-specific peptides with neuroexcitatory activity. Proc Natl Acad Sci USA 95(1):322–327
47. Lin L, Faraco J, Li R, Kadotani H, Rogers W, Lin X, Qiu X, de Jong PJ, Nishino S, Mignot E (1999) The sleep disorder canine narcolepsy is caused by a mutation in the hypocretin (orexin) receptor 2 gene. Cell 98(3):365–376
48. Shiuchi T, Haque MS, Okamoto S, Inoue T, Kageyama H, Lee S, Toda C, Suzuki A, Bachman ES, Kim YB et al (2009) Hypothalamic orexin stimulates feeding-associated glucose utilization in skeletal muscle via sympathetic nervous system. Cell Metab 10(6):466–480

49. Zhang S, Zeitzer JM, Yoshida Y, Wisor JP, Nishino S, Edgar DM, Mignot E (2004) Lesions of the suprachiasmatic nucleus eliminate the daily rhythm of hypocretin-1 release. Sleep 27(4):619–627
50. Scheen AJ, Buxton OM, Jison M, van Reeth O, Leproult R, L'Hermite-Baleriaux M, Van Cauter E (1998) Effects of exercise on neuroendocrine secretions and glucose regulation at different times of day. Am J Physiol 274(6 Pt 1):E1040–E1049
51. Kalsbeek A, Ruiter M, la Fleur SE, van Heijningen C, Bujis RM (2003) The diurnal modulation of hormonal responses in the rat varies with different stimuli. J Neuroendocrinol 15(12):1144–1155
52. Jasper MS, Engeland WC (1994) Splanchnic neural activity modulates ultradian and circadian rhythms in adrenocortical secretion in awake rats. Neuroendocrinology 59(2):97–109
53. Kaneko M, Kaneko K, Shinsako J, Dallman MF (1981) Adrenal sensitivity to adrenocorticotropin varies diurnally. Endocrinology 109(1):70–75
54. Kalsbeek A, van der Spek R, Lei J, Endert E, Buijs RM, Fliers E (2012) Circadian rhythms in the hypothalamo-pituitary-adrenal (HPA) axis. Mol Cell Endocrinol 349(1):20–29
55. Ruiter M, la Fleur SE, van Heijningen C, van der Vleit J, Kalsbeek A, Buijs RM (2003) The daily rhythm in plasma glucagon concentrations in the rat is modulated by the biological clock and by feeding behavior. Diabetes 52(7):1709–1715
56. Dallman MF, Strack AM, Akana SF, Bradbury MJ, Hanson ES, Scribner KA, Smith M (1993) Feast and famine: critical role of glucocorticoids with insulin in daily energy flow. Front Neuroendocrinol 14(4):303–347
57. De Boer SF, Van der Gugten J (1987) Daily variations in plasma noradrenaline, adrenaline and corticosterone concentrations in rats. Physiol Behav 40(3):323–328
58. Kalsbeek A, van der Vliet J, Buijs RM (1996) Decrease of endogenous vasopressin release necessary for expression of the circadian rise in plasma corticosterone: a reverse microdialysis study. J Neuroendocrinol 8(4):299–307
59. Trumper BG, Reschke K, Molling J (1995) Circadian variation of insulin requirement in insulin dependent diabetes mellitus the relationship between circadian change in insulin demand and diurnal patterns of growth hormone, cortisol and glucagon during euglycemia. Horm Metab Res 27(3):141–147
60. van Cauter E (1990) Diurnal and ultradian rhythms in human endocrine function: a minireview. Horm Res 34(2):45–53
61. Bright GM, Melton TW, Rogol AD, Clarke WL (1980) Failure of cortisol blockade to inhibit early morning increases in basal insulin requirements in fasting insulin-dependent diabetics. Diabetes 29(8):662–664
62. Cailotto C, la Fleur SE, van Heijningen C, Wortel J, Kalsbeek A, Feenstra M, Pevet P, Bujis RM (2005) The suprachiasmatic nucleus controls the daily variation of plasma glucose via the autonomic output to the liver: are the clock genes involved? Eur J Neurosci 22(10):2531–2540
63. Clark RG, Chambers G, Lewin J, Robinson IC (1986) Automated repetitive microsampling of blood: growth hormone profiles in conscious male rats. J Endocrinol 111(1):27–35
64. Kimura F, Tsai CW (1984) Ultradian rhythm of growth hormone secretion and sleep in the adult male rat. J Physiol 353:305–315
65. DeFronzo RA (2010) Insulin resistance, lipotoxicity, type 2 diabetes and atherosclerosis: the missing links. The Claude Bernard Lecture 2009. Diabetologia 53(7):1270–1287
66. DeFronzo RA, Simonson D, Ferrannini E (1982) Hepatic and peripheral insulin resistance: a common feature of type 2 (non-insulin-dependent) and type 1 (insulin-dependent) diabetes mellitus. Diabetologia 23(4):313–319
67. Butler AE, Janson J, Soeller WC, Butler PC (2003) Increased beta-cell apoptosis prevents adaptive increase in beta-cell mass in mouse model of type 2 diabetes: evidence for role of islet amyloid formation rather than direct action of amyloid. Diabetes 52(9):2304–2314
68. Perley MJ, Kipnis DM (1967) Plasma insulin responses to oral and intravenous glucose: studies in normal and diabetic subjects. J Clin Invest 46(12):1954–1962
69. Yki-Jarvinen H (1995) Role of insulin resistance in the pathogenesis of NIDDM. Diabetologia 38(12):1378–1388

70. Florez JC (2008) Newly identified loci highlight beta cell dysfunction as a key cause of type 2 diabetes: where are the insulin resistance genes? Diabetologia 51(7):1100–1110
71. Voight BF, Scott LJ, Steinthorsdottir V, Morris AP, Dina C, Welch RP, Zeggini E, Huth C, Aulchenko YS, Thorleifsson G et al (2010) Twelve type 2 diabetes susceptibility loci identified through large-scale association analysis. Nat Genet 42(7):579–589
72. Karlsson B, Knutsson A, Lindahl B (2001) Is there an association between shift work and having a metabolic syndrome? Results from a population based study of 27,485 people. Occup Environ Med 58(11):747–752
73. Kroenke CH, Spiegelman D, Manson J, Schernhammer ES, Colditz GA, Kawachi I (2007) Work characteristics and incidence of type 2 diabetes in women. Am J Epidemiol 165(2):175–183
74. Morikawa Y, Nakagawa H, Miura K, Soyama Y, Ishizaki M, Kido T, Naruse Y, Suwazono Y, Nogawa K (2007) Effect of shift work on body mass index and metabolic parameters. Scand J Work Environ Health 33(1):45–50
75. Scheer FA, Hilton MF, Mantzoros CS, Shea SA (2009) Adverse metabolic and cardiovascular consequences of circadian misalignment. Proc Natl Acad Sci USA 106(11):4453–4458
76. Roenneberg T, Wirz-Justice A, Merrow M (2003) Life between clocks: daily temporal patterns of human chronotypes. J Biol Rhythms 18(1):80–90
77. Roenneberg T, Kuehnle T, Pramstaller PP, Ricken J, Havel M, Guth A, Merrow M (2004) A marker for the end of adolescence. Curr Biol 14(24):R1038–R1039
78. van Cauter E (2011) Sleep disturbances and insulin resistance. Diabet Med 28(12):1455–1462
79. Storch KF, Lipan O, Leykin I, Viswanathan N, Davis FC, Wong WH, Weitz CJ (2002) Extensive and divergent circadian gene expression in liver and heart. Nature 417(6884):78–83
80. Muhlbauer E, Wolgast S, Finckh U, Peschke D, Peschke E (2004) Indication of circadian oscillations in the rat pancreas. FEBS Lett 564(1–2):91–96
81. Zvonic S, Ptitsyn AA, Conrad SA, Scott LK, Floyd ZE, Kilroy G, Wu X, Goh BC, Mynatt RL, Gimble JM (2006) Characterization of peripheral circadian clocks in adipose tissues. Diabetes 55(4):962–970
82. Reppert SM, Weaver DR (2002) Coordination of circadian timing in mammals. Nature 418(6901):935–941
83. Damiola F, Le MN, Preitner N, Kornmann B, Fleury-Olela F, Schibler U (2000) Restricted feeding uncouples circadian oscillators in peripheral tissues from the central pacemaker in the suprachiasmatic nucleus. Genes Dev 14(23):2950–2961
84. Dufour CR, Levasseur MP, Pham NH, Eichner LJ, Wilson BJ, Charest-Marcotte A, Duguay D, Poirier-Heon JF, Cermakian N, Giguere V (2011) Genomic convergence among ERRalpha, PROX1, and BMAL1 in the control of metabolic clock outputs. PLoS Genet 7(6):e1002143
85. Lamia KA, Sachdeva UM, DiTacchio L, Williams EC, Alvarez JG, Egan DF, Vasquez DS, Juguilon H, Panda S, Shaw RJ et al (2009) AMPK regulates the circadian clock by cryptochrome phosphorylation and degradation. Science 326(5951):437–440
86. Liu C, Li S, Liu T, Borjigin J, Lin JD (2007) Transcriptional coactivator PGC-1alpha integrates the mammalian clock and energy metabolism. Nature 447(7143):477–481
87. Hsieh MC, Yang SC, Tseng HL, Hwang LL, Chen CT, Shieh KR (2010) Abnormal expressions of circadian-clock and circadian clock-controlled genes in the livers and kidneys of long-term, high-fat-diet-treated mice. Int J Obes (Lond) 34(2):227–239
88. Barnea M, Madar Z, Froy O (2010) High-fat diet followed by fasting disrupts circadian expression of adiponectin signaling pathway in muscle and adipose tissue. Obesity (Silver Spring) 18(2):230–238
89. Barnea M, Madar Z, Froy O (2009) High-fat diet delays and fasting advances the circadian expression of adiponectin signaling components in mouse liver. Endocrinology 150(1):161–168
90. Kohsaka A, Laposky AD, Ramsey KM, Estrada C, Joshu C, Kobayashi Y, Turek FW, Bass J (2007) High-fat diet disrupts behavioral and molecular circadian rhythms in mice. Cell Metab 6(5):414–421

91. Sutton GM, Centanni AV, Butler AA (2010) Protein malnutrition during pregnancy in C57BL/6J mice results in offspring with altered circadian physiology before obesity. Endocrinology 151(4):1570–1580
92. Ando H, Ushijima K, Yanagihara H, Hayashi Y, Takamura T, Kaneko S, Fujimura A (2009) Clock gene expression in the liver and adipose tissues of non-obese type 2 diabetic Goto-Kakizaki rats. Clin Exp Hypertens 31(3):201–207
93. Otway DT, Mantele S, Bretschneider S, Wright J, Trayhurn P, Skene DJ, Robertson MD, Johnston JD (2011) Rhythmic diurnal gene expression in human adipose tissue from individuals who are lean, overweight, and type 2 diabetic. Diabetes 60(5):1577–1581
94. Turek FW, Joshu C, Kohsaka A, Lin E, Ivanova G, McDearmon E, Laposky A, Losee-Olson S, Easton A, Jensen DR et al (2005) Obesity and metabolic syndrome in circadian Clock mutant mice. Science 308(5724):1043–1045
95. Marcheva B, Ramsey KM, Buhr ED, Kobayashi Y, Su H, Ko CH, Ivanova G, Omura C, Mo S, Vitaterna MH et al (2010) Disruption of the clock components CLOCK and BMAL1 leads to hypoinsulinaemia and diabetes. Nature 466(7306):627–631
96. Oishi K, Ohkura N, Wakabayashi M, Shirai H, Sato K, Matsuda J, Atsumi G, Ishida N (2006) CLOCK is involved in obesity-induced disordered fibrinolysis in ob/ob mice by regulating PAI-1 gene expression. J Thromb Haemost 4(8):1774–1780
97. Sadacca LA, Lamia KA, DeLemos AS, Blum B, Weitz CJ (2011) An intrinsic circadian clock of the pancreas is required for normal insulin release and glucose homeostasis in mice. Diabetologia 54(1):120–124
98. Vieira E, Marroqui L, Batista TM, Caballero-Garrido E, Carneiro EM, Boschero AC, Nadal A, Quesada I (2012) The clock gene Rev-erbalpha regulates pancreatic beta-cell function: modulation by leptin and high-fat diet. Endocrinology 153(2):592–601
99. Dupuis J, Langenberg C, Prokopenko I, Saxena R, Soranzo N, Jackson AU, Wheeler E, Glazer NL, Bouatia-Naji N, Gloyn AL et al (2010) New genetic loci implicated in fasting glucose homeostasis and their impact on type 2 diabetes risk. Nat Genet 42(2):105–116
100. Liu C, Li H, Qi L, Loos RJ, Qi Q, Lu L, Gan W, Lin X (2011) Variants in GLIS3 and CRY2 are associated with type 2 diabetes and impaired fasting glucose in Chinese Hans. PLoS One 6(6):e21464
101. Oishi K, Atsumi G, Sugiyama S, Kodomari I, Kasamatsu M, Machida K, Ishida N (2006) Disrupted fat absorption attenuates obesity induced by a high-fat diet in Clock mutant mice. FEBS Lett 580(1):127–130
102. Kennaway DJ, Owens JA, Voultsios A, Boden MJ, Varcoe TJ (2007) Metabolic homeostasis in mice with disrupted Clock gene expression in peripheral tissues. Am J Physiol Regul Integr Comp Physiol 293(4):R1528–R1537
103. Toye AA, Lippiat JD, Proks P, Shimomura K, Bentley L, Hugill A, Mijat V, Goldsworthy M, Moir L, Haynes A et al (2005) A genetic and physiological study of impaired glucose homeostasis control in C57BL/6J mice. Diabetologia 48(4):675–686
104. Bunger MK, Wilsbacher LD, Moran SM, Clendenin C, Radcliffe LA, Hogenesch JB, Simon MC, Takahashi JS, Bradfield CA (2000) Mop3 is an essential component of the master circadian pacemaker in mammals. Cell 103(7):1009–1017
105. Dallmann R, Touma C, Palme R, Albrecht U, Steinlechner S (2006) Impaired daily glucocorticoid rhythm in Per1 (Brd) mice. J Comp Physiol A Neuroethol Sens Neural Behav Physiol 192(7):769–775
106. Raspe E, Mautino G, Duval C, Fontaine C, Duez H, Barbier O, Monte D, Fruchart J, Fruchart JC, Staels B (2002) Transcriptional regulation of human Rev-erbalpha gene expression by the orphan nuclear receptor retinoic acid-related orphan receptor alpha. J Biol Chem 277(51):49275–49281
107. Kornmann B, Schaad O, Bujard H, Takahashi JS, Schibler U (2007) System-driven and oscillator-dependent circadian transcription in mice with a conditionally active liver clock. PLoS Biol 5(2):e34

108. Bray MS, Young ME (2009) The role of cell-specific circadian clocks in metabolism and disease. Obes Rev 10(Suppl 2):6–13
109. Lamia KA, Storch KF, Weitz CJ (2008) Physiological significance of a peripheral tissue circadian clock. Proc Natl Acad Sci USA 105(39):15172–15177
110. Yin L, Wu N, Curtin JC, Qatanani M, Szwergold NR, Reid RA, Waitt GM, Parks DJ, Pearce KH, Wisely GB et al (2007) Rev-erbalpha, a heme sensor that coordinates metabolic and circadian pathways. Science 318(5857):1786–1789
111. Zhang EE, Liu Y, Dentin R, Pongsawakul PY, Liu AC, Hirota T, Nusinow DA, Sun X, Landais S, Kodama Y et al (2010) Cryptochrome mediates circadian regulation of cAMP signaling and hepatic gluconeogenesis. Nat Med 16(10):1152–1156
112. Kalsbeek A, Yi CX, la Fleur SE, Fliers E (2010) The hypothalamic clock and its control of glucose homeostasis. Trends Endocrinol Metab 21(7):402410
113. Green CB, Takahashi JS, Bass J (2008) The meter of metabolism. Cell 134(5):728–742.
114. Nagai K and Nakagawa H (1992) Central Regulation of Energy Metabolism with special reference to circadian rhythm. CRC Press, Boca Raton, Florida

Chapter 8
Genetics in Chronobiology and Obesity

Marta Garaulet and Jose M. Ordovás

Abstract Genetics is behind our circadian machinery. Some of our chronobiological characteristics could be influenced by genes. Different psychological traits such as depression, bipolar disorders, anxiety and seasonal variations of mood are intrinsically connected to chronobiology through different genetic variants. Moreover, sleep disorders or short sleep duration, are both associated to several polymorphisms connected to obesity. In this regards, one of the most outstanding SNPs is the *CLOCK* 3111TC SNP which is significantly associated to short sleep duration, eveningness, several psychological traits and obesity. This SNP has been also related to a reduction in weight loss effectiveness in patients submitted to a behavioral treatment of obesity. Ghrelin, eveningness, and a lack of compliance to the Mediterranean diet habits, could be behind these results. Apart from *CLOCK* SNPs, others genetic variants in several clock genes such as *PERIOD* or *BMAL1* are also connected to obesity. The novel knowledge achieved in the circadian epigenome could give us new answers to the connections among genetics, circadian rhythmicity and obesity.

M. Garaulet, Ph.D. (✉)
Department of Physiology, University of Murcia, Murcia, Spain

J.M. Ordovás, Ph.D.
Functional Genomics Core, Nutrition and Genomics Laboratory, Jean Mayer USDA, HNRCA at Tufts University, 711 Washington Street, Boston, MA, 02111-1524, USA

The Department of Epidemiology and Population Genetics, Centro Nacional Investigaciones Cardiovasculares (CNIC), Instituto Madrileño de Estudios Avanzados en Alimentación (IMDEA-FOOD), Madrid, Spain
e-mail: jose.ordovas@tufts.edu

M. Garaulet and J.M. Ordovás (eds.), *Chronobiology and Obesity*,
DOI 10.1007/978-1-4614-5082-5_8,

Abbreviations

SNP	Single nucleotide polymorphism
CLOCK	Circadian locomotor output cycles kaput
PER	Period homolog 2 (Drosophila)
BMAL1 or ARNTL or MOP3	Aryl hydrocarbon receptor nuclear translocator-like
HGP	Human genome project
DNA	Deoxyribonucleic acid
RNA	Ribonucleic acid
mRNA	Messenger ribonucleic acid
GH	Growth hormone
SCN	Suprachiasmatic nucleus
MetS	Metabolic syndrome
MD	Mood disorders
OMIM	Online Mendelian inheritance in man
CRY	Cryptochrome
REV-ERBα	Nuclear receptor Rev-ErbA-alpha
SIRT	Sirtuin
RORA or NR1D1	RAR-related orphan receptor A
VIP	Vasoactive intestinal polypeptide
ROR1	Receptor tyrosine kinase-like orphan receptor 1
PLCB1	Phospholipase C, beta 1
OSAS	Obstructive sleep apnea syndrome
MTNR1A	Melatonin receptor 1A
MTNR1B	Melatonin receptor 1B
GWAS	Genome-wide association studies
NFATC2	Nuclear factor of activated T cells 2
SCP2	Sterol carrier protein 2
CACNA1C	Calcium channel, voltage-dependent, L type, alpha 1C subunit
TCRA	T cell receptor alpha chain
POLE	Polymerase (DNA directed), epsilon
FAM3D	Family with sequence similarity 3, member D
ABCC9	ATP-binding cassette, sub-family C (CFTR/MRP), member 9
SUR2	Potential sterol desaturase similar to *S. cerevisiae*
HLA	Human leukocyte antigen
DQB1	Major histocompatibility complex, class II, DQ beta 1
PSD	Partial sleep deprivation
NPAS2	Neuronal PAS domain protein 2
APSS	Associated Professional Sleep Societies LLC
FTO	Fat mass and obesity associated
HOMA-IR	Homeostasis model assessment- insulin resistance

TMEM18	Transmembrane protein 18
NRXN3	Neurexin 3
BMI	Body mass index
GOLDN	Genetics of Lipids Lowering Drugs and Diet Network
FAs	Fatty acids
MUFA	Monounsaturated fatty acid
SFA	Saturated fatty acid
MCP1	Monocyte chemoattractant protein 1
IL-6	Interleukin 6
PTMs	Post translational modifications
HAT	Histone acetile transferase
BMI	Body mass index
SAT	Saturated fatty acids
MUFA	Monounsaturated fatty acids

The Human Genome Revolution: Feast or Famine?

The official presentation of the final Human Genome Project (HGP) was carried out in April, 2003 coinciding with the 50th anniversary of the seminal publication of the structure of the double helix of DNA for Watson and Crick. Though, already in the year 2001, the drafts of the human genome were published in two of the most prestigious scientific journals, Nature and Science. Shortly after, in 2007, the first two individual sequences were accomplished, one of them, belonging to James Watson was announced in Nature News on June 1 and the other, belonging to Craig Venter was published in PLOS Biology on September 4. Since then, thousands of genomes have been completed at incredibly faster speed and lower cost. The ultimate goal will be to use this information to develop new ways to treat, cure, or even prevent the thousands of diseases that afflict humankind.

However, the HGP has failed so far to produce the announced health revolution that some scientists, and other members of the society, had promised in a relatively short time. Indeed, the concept of "personalized medicine" that was supposed to provide with and make use of tests informing a person's risk for heart disease, cancer and other common illnesses, is demonstrating to be not as simple as some may have anticipated. Conversely, we are continuously learning about previously unknown levels of biological complexity, including not only the polygenic nature of all common diseases, but also the role of multiple epigenetic mechanisms that are heavily influenced by environmental factors.

Therefore, our current view shows that most common diseases arises from regulatory or structural dysfunctions of multiple genes that interact with a myriad of environmental and behavioral factors, including our own microbiota.

The Birth of New Genetic-Related Sciences: Nutrigenetics, Nutrigenomics and Epigenetics

The new perspective generated by the HGP and the ensuing research has fostered and intensified some areas of more specialized genetic research aimed to advance our knowledge and to translate this knowledge into practical applications

Among these new areas, we could highlight *Nutritional Genomic* which focuses on the relationship between human genome, nutrition and health that can be divided into two subspecialties:

(a) *Nutrigenomic* which studies the effect of nutrients on health through altering genome, proteome, metabolome and the resulting changes in physiology. For example, recently, it has been discovered that the health effects of food compounds are related mostly to specific interactions on molecular level, i.e., dietary constituents participate in the regulation of gene expression.
(b) *Nutrigenetic* which studies the effect of genetic variations on the interaction between diet and health with implications to susceptible subgroups. The genetic variation or SNPs (Single Nucleotide Polymorphisms) are changes in only one nucleotide that is the most frequent cause of the different responses to a diet or a drug. More specifically, Nutrigenetic studies how individual differences in genes influence the body's response to diet and nutrition. This necessitates the identification of gene variants associated with differential responses to nutrients and with higher susceptibility to diet-related diseases. The ultimate goal of nutrigenetics is to provide nutritional recommendations for individuals in what is known as personalized or individualized nutrition.

Another blossoming area relates to *Epigenetics* which is the study of heritable changes in gene expression or cellular phenotype caused by mechanisms other than changes in the underlying DNA sequence—hence the name epi- (Greek: επ - over, above, outer) -genetics. It refers to functionally relevant modifications to the genome that do not involve a change in the nucleotide sequence. Examples of such changes are DNA methylation and histone modification, both of which serve to regulate gene expression without altering the underlying DNA sequence. Moreover, as recently as 2011, it has been demonstrated that the methylation of mRNA has a critical role in human energy homeostasis, opening the related field of RNA epigenetics.

How Could Genetics Be Associated to Chronobiology? A Little Bit of History

The capacity to undergo rhythmic oscillations is a characteristic intrinsic to living matter. A fundamental statement of chronobiology states "many rhythms persist even in complete isolation from the major known environmental cycles." This concept supports that natural rhythms can exist independently of the periods defined by geophysical cycles; this means that living matter has its own time, i.e., the "biological time." In this sense, it has been hypothesized the existence of a Chronome within the Genome [1].

Over the past two decades, biochemical, genetic, and molecular studies have been making substantial advances towards the elucidation of the molecular bases of rhythmicity in living things. Riding on the wave generated by the seminal studies in the 1970s focusing on in the circadian variability of hormones such as cortisol, melatonin or growth hormone (GH), or those related to the discovering and description of the physiological bases of the suprachiasmatic nucleus (SCN), current chronobiology has dramatically evolved thanks to the new genetic and molecular biology techniques.

A major stride in understanding the molecular basis of circadian rhythms was the identification by Konopa and Benzer in 1971 of a chromosomal region controlling the period of eclosion time in Drosophila, followed by the cloning of the first clock genes in *Drosophila melanogaster* in 1984. Today, thanks to these molecular techniques, we are able to study the expression of the known clock genes implicated in the circadian machinery. We already know that, in mammals, the core components of the clock molecular machinery operate in almost all cells of the body through a complex network of transcriptional-translation loops and modulate the expression of specific target genes and their products to oscillate in 24-h rhythm [2].

Nowadays, experimental models are allowing us to assess clock genes expression not only in the living animal but also outside of the body (in vitro techniques) and we are also able to analyze the 24 h fluctuations in gene expression and to assess the presence or absence of a peripheral clock in the different organs and tissues (see Chap. 2). Moreover, we can use experimental models to turn on and off specific components of the clock machinery to identify its effects on metabolic and disease phenotypes. From the genetic epidemiology point of view, the study of single nucleotide polymorphisms (SNPs), is contributing to the identification of the genetic background of chronotypes (morningness or eveningness), sleep alterations, or seasonal mood disorders.

More recently, epigenetic and nutrigenetic approaches have also been allowing us to study new interactions and layers of complexity that may have a significant impact on chronobiology as well as pathophysiology

Finally, the technological power of other "-omics" (i.e., metabolomics, proteomics) is becoming essential to our ability to "put-it-all-together" and we are fast learning about the timing of different metabolites such as aminoacids, lipids, xenobiotic, etc. in the liver in mice [3], and in plasma and saliva in humans [4], allowing us to achieve a more complete and refined knowledge of the circadian rhythm and its physiological effects. These advances have given to the science of chronobiology a renewed stimulus that makes this science increasingly robust and attractive.

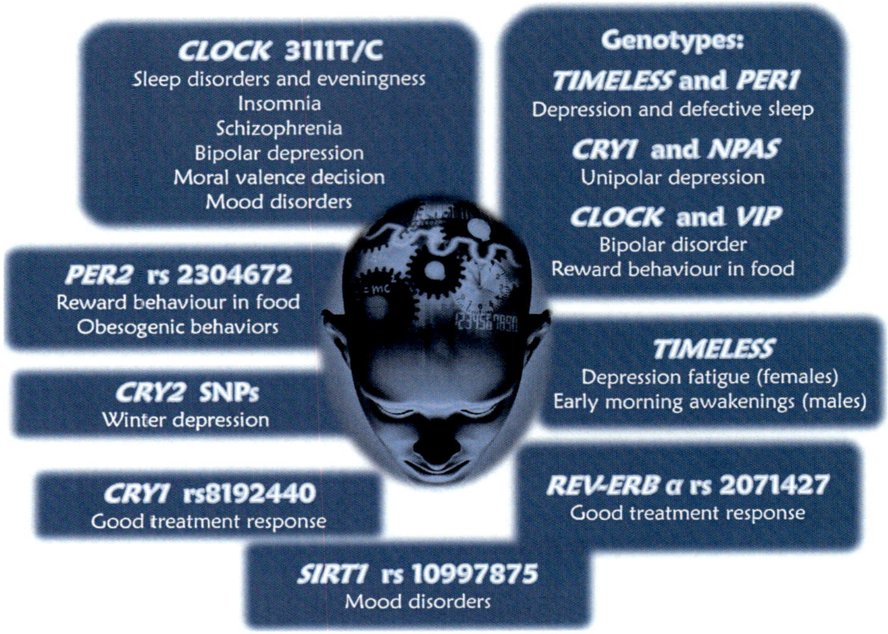

Fig. 8.1 Main genetic variants related to psychological traits

Genetic, Chronotype, and Psychological Traits

As shown in other chapters, there are individual differences in preferred times for activity and rest. The characteristics that differentiate "owls" from "larks" may reside primarily on their circadian systems, and presumably genetics prevail over environmental factors.

Initially, the emphasis of chronobiology genetics was placed on the study of the association between chronotype and different psychological illnesses such as depression, anxiety, or bipolar disorder. Other mood features that have been associated with clock genes are stress, seasonality, and personality traits related to the chronotype such as the morningness/eveningness profile.

Taking into account that obesity is related to behavior and also with personality traits, and considering that the chronotype is behind some of these relationships, an obvious step was to link these psychological-related SNPs with obesity-related traits. Proof of this hypothesis is the case of circadian locomotor output cycles kaput (*CLOCK*) and *PERIOD2* (*PER2*) SNPs that were firstly associated with mood disorders and then to obesity and Metabolic Syndrome (MetS) (Fig. 8.1).

The initial studies of clock genes and their association with psychological traits come from the observation that patients with Mood Disorders (MD) commonly show biological rhythm-related symptoms, such as characteristic disturbances in the sleep/ wake cycle, diurnal mood changes, and a periodic pattern of symptom recurrence and remission [5]. In addition, alterations in the circadian pattern of core body temperature and neuroendocrine secretion have been also documented in psychological illnesses. [6] On the other hand, mood-stabilizing drug such lithium or antidepressants such as fluoxetin [7] are known to modulate circadian rhythms [8].

Probably the most studied of the circadian genes has been the *CLOCK* gene. The CLOCK locus is located on the long arm of chromosome 4q12 (Online Mendelian Inheritance in Man (OMIM) *601851, 25 exons in the genomic region spanning 115.138 kb). The interest of this gene is that its translation product is involved in the transcriptional regulation of many circadian output genes and in the core circadian clock.

A common 3111T→C SNP at the 3′-untranslated region of *CLOCK* gene contributes to our ability to classify people as "larks" or "owls." Thus, individuals carrying the 3111C allele tend to define themselves as nocturnal more often than homozygotes for 3111T allele.

Moreover, different psychological traits have been related to this SNP. For example, it has been found an interesting association between *CLOCK* 3111T/C and *attention deficit hyperactivity disorder* [9], psychological related *insomnia* [10], *Schizophrenia* [11] *bipolar depression* [12], *moral valence decision* [13] and *mood disorders* [14]. The association with all these psychological illnesses is related to the fact that minor allele carriers of *CLOCK* 3111TC display sleep disorders and eveningness [15], characteristics that, in addition, make these subjects susceptible to obesity.

Other clock genes have been also related to mood disorders or behaviors associated with obesity. This is the case of *PERIOD2* (*PER2*). A *PER2* SNP (rs2304672) has been shown to moderate *circadian-relevant reward circuitry* activity in adolescents. *Reward behavior* in animals is highly related to food intake and is influenced by circadian genes, including clock-pathway genes such as *PER2*. Several forms of psychiatric illness are associated with both altered reward function and disturbances in circadian function. Associations among circadian genes function in neural reward circuits, and circadian-influenced behavior could be important in obesity. Indeed, in a further work we have related this SNP with attrition during a weight loss treatment and with obesogenic behaviors such as stress with dieting, snacking, or eating when bored.

Cryptochrome genes (*CRY1 and CRY2*) code for the two cryptochrome proteins CRY1 and CRY2 act as light-independent inhibitors of CLOCK-BMAL1 components of the circadian clock and specifically, CRY2 participates in the regulation of the evening oscillator. This is of interest in mood disorders where a deficient switch from evening to morning oscillators has been postulated [5]. Indeed, in *depressed bipolar* patients, levels of *CRY2* mRNA are decreased and there is no response following sleep deprivation. These investigators have shown that genetic variation at the *CRY2* gene was significantly associated with winter depression in two independent population-based samples from Sweden and Finland [16].

Other clock genes have been also related to different behaviors or mood disorders. A functional polymorphism in the *REV-ERBα* (rs2071427) locus and a second variant in *CRY1* (rs8192440), both were nominally associated with positive response to psychological treatment [17]. Moreover, several recent investigations have implicated *SIRT1* in the regulation of the circadian system in combination with the traditional circadian clock genes, and with the dopaminergic pathway. Therefore, the *SIRT1* locus has been added to the rank of candidate genes involved in mood disorders and this is supported by reported associations between the SIRT1 rs10997875 SNP and mood disorders [18].

Other authors have found significant associations between *TIMELESS* variants and depression with *fatigue* in *females*, and association to depression with *early morning awakening* in *males* [19]. Indeed, there was a significant interaction of gender and *TIMELESS*. Authors also obtained supported evidence for involvement of *TIMELESS* in sleeping problems in an independent set of control individuals with seasonal changes in mood, sleep duration, energy level and social activity in females and with early morning awakening or fatigue in males.

There was also some evidence of interaction between *TIMELESS* and *PER1* in females as well as between *TIMELESS* and *ARNTL*, *RORA*, or *NR1D1* in males. These findings support a connection between circadian genes and gender-dependent depression and defective sleep [19]. Other authors have found a differential association of circadian genes with mood disorders: *CRY1* and *NPAS2* are associated with *unipolar major depression* and *CLOCK* and *VIP* with *bipolar disorder* [5]. All these genetic variants could be important in the pathophysiology of obesity, taking into account that many of these alterations are related to food intake and emotions, both aspects highly associated to obesity and weight loss.

Genetics in Sleep Disorders

Other candidate SNPs connected to obesity could be those associated to sleep disorders. The contribution of genes, environment, and gene–environment interactions to sleep disorders is increasingly recognized. Well-documented familial and twin sleep disorder studies suggest an important influence of genetic factors. Most sleep disorders are complex in terms of their genetic susceptibility together with the variable manifestation of the phenotype even within the same family. Recent linkage, genome-wide and candidate gene association studies resulted in the identification of gene mutations, gene localizations, or evidence for susceptibility genes and/or loci in several sleep disorders [20].

One common sleep disorder in the current society is insomnia which is reported to chronically affect 10–15% of the adult population. However, very little is known about the genetics and metabolism of insomnia. A study performed in 10,038 Korean subjects [21] showed that about 16.5% reported insomnia and displayed distinct metabolic changes reflecting an increase in insulin secretion, a higher risk of diabetes, and disrupted calcium signaling. Insomnia-associated genotypic differences were highly concentrated within genes involved in neural function. The most significant SNPs resided in ROR1 and PLCB1 genes known to be involved in bipolar disorder and schizophrenia, respectively [21].

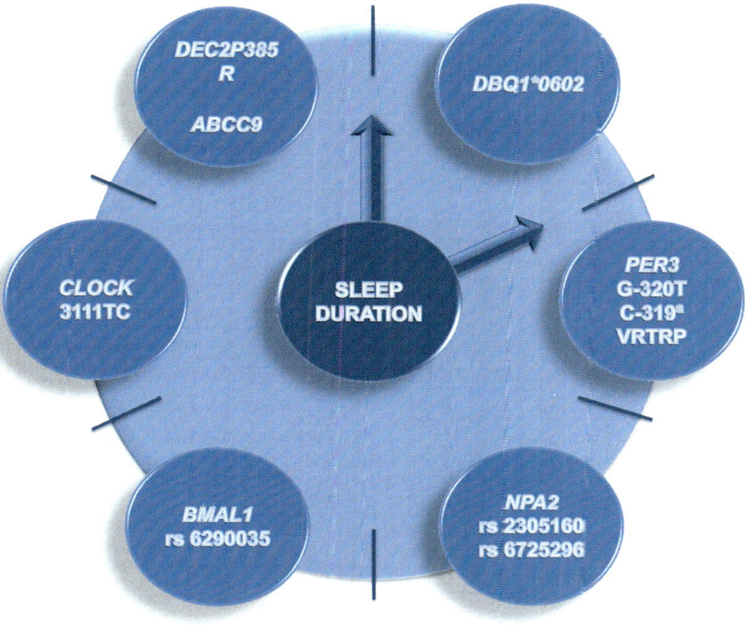

Fig. 8.2 Genetics of sleep duration

Another particular example of sleep disorder highly related to obesity is *Obstructive Sleep Apnea Syndrome* (*OSAS*), and it has been postulated that genetic variants in IL-6 could modify susceptibility to OSAS [22, 23]. They could also explain fatigue in different pathologies such as cancer [24] and could also link sleep with obesity and metabolic syndrome alterations, through inflammation. Others loci have been proposed in relation to sleep disorders; including serotonin receptors SNPs; β2-adrenergic receptor that are related with nocturnal asthma [25] and with nocturnal blood pressure dipping status [26] and the prepro-orexin gene polymorphism g1182C>T which associates with obstructive sleep apnea/hypopnea syndrome [27]. Particular relevant are the studies in SNPs in the promoter region of the melatonin receptor genes (*MTNR1A* and *MTNR1B*) [28] but their role in sleep disorders and its association with obesity and MetS still remain controversial [29].

Narcolepsy is a sleep disorder characterized by excessive daytime sleepiness, cataplexy, and a pathological manifestation of rapid eye movement during sleep. Narcoleptic pathogenesis is triggered by both genetic and environmental factors. Genome-wide association studies (GWAS) have identified over 200 new genetic factors. Following replication in 222 narcoleptic patients and 380 controls, six genes, *NFATC2*, *SCP2*, *CACNA1C*, *TCRA*, *POLE*, and *FAM3D*, remained associated with narcolepsy [30]. One can speculate that some of these loci could contribute to our understanding of the relation between sleep disorders, obesity and other metabolic diseases.

Genetics in Sleep Duration

As has been referred in Chap. 7, sleep duration is related to obesity and genetic variability could be implicated in this relationship. Indeed, in the last years a number of SNPs have been related to sleep duration and some of them are being further studied in obesity (Fig. 8.2).

One of the most promising findings relates to the identification of a P385R mutation in a transcriptional repressor (hDEC2-P385R) that was associated with a human short sleep phenotype. The habitual self-reported total sleep time per 24-h day was much shorter in mutation carriers (average 6.25 h) compared with the noncarriers (average 8.06 h). Thus, they represent "natural short sleepers" who routinely sleep less than individuals with familial advanced sleep-phase syndrome (FASPS) or general controls. In order to proof the functionality of this variant, a wild-type or a P385R mDec2 construct was used in a luciferase assay, and the results indicated that the activity profiles and sleep recordings of transgenic mice carrying this mutation showed increased vigilance time and less sleep time than control mice in a zeitgeber time- and sleep deprivation-dependent manner [31]. However, although short sleep has been related to obesity, to our knowledge there is no such study has been carried for DEC2-P385R

One particular example that could link *sleep duration and obesity* is *ABCC9*, for which functional studies have shown that the protein plays a role in the pathogenesis of heart disease and diabetes and which also influences the duration of sleep in humans. The specific variant (rs11046205) was first discovered in a GWAS that investigated sleep patterns. More than 4,000 people from seven European populations, from countries as diverse as Estonia and Italy, took part in the project, and filled out a questionnaire designed to assess their sleeping habits. Analysis of the genetic and behavioral data revealed that individuals who had two copies of the minor ABCC9 rs11046205 A allele generally slept for a significantly shorter period (~5% less) in an undisturbed environment than did persons homozygotes for the common G allele [32]. The gene *ABCC9* codes for the protein SUR2, which forms the regulatory component of a potassium channel in the cell membrane. This ion channel acts as a sensor of energy metabolism in the cell.

Other authors have proposed the human leukocyte antigen *(HLA) DQB1*0602* allele may represent a genetic biomarker for predicting individual differences in both basal and sleep loss conditions. The influence of the *DQB1*0602* allele on sleep homeostatic and neurobehavioral responses has been examined in chronic partial sleep deprivation (PSD) [33] and in healthy subjects.

An interesting study that could further connect sleep with obesity is the one performed in *PERIOD3* (*PER3*) by Archer et al. [34] who screened the *PER3* promoter for polymorphisms and investigated the phenotypic associations of these polymorphisms with diurnal preference, delayed sleep phase disorder/syndrome and their effects on reporter gene expression. Authors demonstrated that SNPs G-320T, C-319A, occurred more frequently in sleep phase disorder subjects compared to morning or evening type and that polymorphisms in the *PER3* promoter could affect its expression, leading to potential differences in the observed functions of *PER3* [34].

Another study, in which the population was stratified according to homozygosity for a variable-number (4 or 5) tandem-repeat polymorphism in the coding region of the clock gene *PER3,* indicated that this polymorphism conferred vulnerability to sleep loss and circadian misalignment through its effects on sleep homeostasis. Indeed, in the vulnerable genotype, activation in a posterior prefrontal area was already reduced when comparing the evening to the morning during a normal sleep–wake cycle. Furthermore, in the morning after a night of sleep loss, widespread reductions in activation in prefrontal, temporal, parietal and occipital areas were observed in this genotype [35]. It remains to be investigated whether this different vulnerability to sleep loss is also related to food intake, obesity or MetS characteristics.

Other clock genes are *NPAS2* and *BMAL 1* (or ARNTL) both important genes in the positive control of the clock machinery. In a study performed to assess seasonality and fertility in adults living in Finland [36] it has been concluded that *NPAS2* rs2305160 A allele carriers had lower Global Seasonality Scores, a sum score of six items, i.e., seasonal variation of sleep length, social activity, mood, weight, appetite, and energy level. Furthermore, carriers of the NPAS2 rs6725296 A allele had greater loadings on the metabolic factor (weight and appetite) of the global seasonality score, whereas individuals with ARNTL rs6290035 TT genotype experienced less seasonal variation of energy level. Considering these interesting results, further studies should get into these particular SNPs to assess their potential association with obesity.

Another obvious candidate gene potentially bridging sleep and obesity disorders is *CLOCK*. Along these lines, we have proposed that the association of the *CLOCK* 3111T/C SNP with obesity or weight loss could be mediated by sleep reduction and by ghrelin, connecting sleep, energy intake, and obesity genetics [15].

An association between variants of the human *CLOCK* gene and sleep duration has been reported in two independent populations [15]. In this study, sleep duration was assessed in Central Europe, Estonia, and South Tyrol ($n \sim 77{,}000$) with the Munich ChronoType Questionnaire. A follow-up association study was conducted with subjects from South Tyrol with short (<7 h) and long (>8.5 h) sleepers from Estonia (confirmation sample; $n = 1{,}011$). One hundred ninety-four SNPs covering 19 candidate clock genes were genotyped. From all these SNPs, single and multimarker associations were found within a CLOCK gene intronic region (rs12649507 and rs11932595). Moreover, in the meta-analysis between South Tyrol and Estonia association signals, rs12649507 remained significant.

Although there are multiple SNPs related to sleep that could consequently be associated with obesity, we need to consider that *sleeping less at night may increase the expression of genetic risks for obesity*, while getting enough sleep may suppress genetic influences on body weight [37]. Indeed, in this study, authors indicate that the heritability of BMI when sleep duration equaled 7 h was more than twice as large as the heritability of BMI when sleep duration equaled 9 h.

According to Nathaniel Watson, "there appears to be something about short sleep that creates a permissive environment for expression of obesity-related genes"[36]. Consistent with this notion, a recent work [38], has demonstrated in adolescents that carriers of the TT allele (Risk allele) for one of the most important SNPs related to obesity the FTO SNP, exhibited associations between decreasing sleep duration and increasing BMI, waist circumference, visceral fat and Homeostasis model assessment-insulin resistance (HOMA-IR) (all $P<0.05$). Similar associations were observed in children with risk alleles (but not in those without risk alleles) for the *TMEM18* and *NRXN3* SNPs. On average, 2 h of sleep less per night was associated with an increase in body mass index (BMI) and with more waist circumference in genetically susceptible children.

Genetics in Chronobiology and Obesity

The great inter-individual differences observed in chronotype, responses to sleep curtailment, and association with obesity, point to an underlying genetic component, and some limited data suggest that common genetic polymorphisms involved in circadian regulation may underlie these large phenotypic differences. However, much more understanding is needed about the genetic basis of differential vulnerability in healthy subjects undergoing sleep deprivation, shift work, constant light exposure and snacking and the effects on obesity.

Animal Model

The current knowledge in the association between chronodisruption and obesity initially came from the studies in genetic mouse models of obesity which have demonstrated disrupted circadian sleep–wake patterns. Leptin-deficient ob/ob mouse and leptin-receptor db/db mouse show increased non-REM sleep time, decreased sleep consolidation, decreased locomotory activity, and a smaller compensatory rebound response to acute sleep deprivation [39, 40]. However, it was not until Turek et al. [41], study had been performed that the evidence of a molecular interaction between clock genes and obesity characteristics came out. This study revealed that mice with disruption of the Clock gene were prone to develop a phenotype resembling obesity and Mets. Previously, in 2004, Rudic et al. [42] already showed that mutations in *Clock* and *Bmal1* were associated with impaired glucose tolerance and more recently it has been also demonstrated that these mutant mice modified circadian variation in glucose and triglyceride [43].

A more recent article has shown that deficiency of, *Bmal1*, induces dyslipidemia and ectopic fat formation, indicating that *Bmal1* is involved in the utilization of fat as an energy source. Indeed, lack of *Bmal1* reduced the capacity of fat storage in adipose tissue, resulting in an increase in the levels of circulating fatty acids, including triglycerides, free fatty acids, and cholesterol. Elevation of the circulating fatty acids level induced the formation of ectopic fat in the liver and skeletal muscle [44].

Previous works have indicated that *Bmal*-deficient mice are characterized by having a greater amount of adipose tissue as compared to mice without this deficiency. However, there are conflicting studies suggesting that *Bmal1* plays a significant role in the regulation of adipose tissue differentiation, and also in lipogenesis of mature adipocytes.

Another interesting gene related to MetS is *Cry*. Indeed, transgenic mice overexpressing mutant *Cry1* develop symptoms of MetS, including polydipsia, polyuria, and hyperglycemia. A question that arises from these animal genetic models concerns whether the metabolic phenotypes are due to the disruption of the core clock mechanism itself, or whether they are secondary to altered feeding patterns present in these mice.

Human Genetics Studies

Given the above evidence in experimental models and the results of emerging epidemiological studies showing that alteration in circadian rhythmicity results in pathophysiological changes resembling MetS, there is active research examining the role of clock-related genes in human obesity and MetS alterations. So far, from the multiple genes within the clock machinery, *CLOCK* and *BMAL1* and *PERIOD* are the genes most frequently related to obesity (Table 8.1) and indication of the long way ahead.

CLOCK Gene and OBESITY

Since 2008, following the studies of Sookian et al. [45] in an Argentinean population, and those carried out by Scott's [46] group in an European population, it has been known that different *CLOCK* variants are associated with obesity and MetS, particularly with abdominal obesity [47].

Subsequently, our group has replicated these data in a North American white sample of 540 men and 560 women who participated in the Genetics of Lipids Lowering Drugs and Diet Network (GOLDN) [47] In this population, *CLOCK* SNPs (rs3749474, rs4580704, and rs1801260 (3111TC)) were associated with body mass index (BMI), energy intake, and different variables related to obesity [48]. In fact, our results showed that individuals carrying the minor alleles ate more, slept less, ate more fat, and were more obese (Fig. 8.3). They particularly showed greater abdominal obesity, characterized as being the type of obesity with the greatest metabolic risk. Some of these associations may be functionally explained, such as the *CLOCK* polymorphism rs3749474, potentially leading to a change in mRNA structure that may affect its expression.

Table 8.1 Associations studies of Clock genes polymorphisms, obesity, and Metabolic Syndrome traits

Gene	SNP reference, genotype and Haplotype	Minor allele or Haplotype Frequency (%)	Population	Trait	Outcome	Reference
CLOCK	Haplotypes: (rs4864548/rs3736544/rs1801260) CAT TGT CGC	CAT:31 TGT:33 CGC:28	537 men from White European population.	Metabolic syndrome.	No significant associations between any SNPs and the MetS subcomponents. The CGC haplotype may be protective for the development of obesity.	[46]
	rs1554483 C/G rs6843722 A/C rs6850524 G/C rs4864548 G/A Haplotype: rs1554483/rs4864548 CG/GA	G: 43 C: 41 C: 34 A: 43	Lean ($n = 715$) and overweight or obese ($n = 391$).	Overweight and obesity.	Four tSNP showed significant differences between lean and overweight/obese. No association was observed with MetS subcomponents. rs1554483G/rs4864548A haplotype was associated with a 1.8-fold risk of obesity.	[45]
	rs1801260 3111T/C	C:53	284 Caucasian subjects including 92 normal weight and 192 overweight/obese.	Binge eating disorders and BMI.	Genotype and allele frequencies did not significantly differ between normal weight and obese patients with and/or without BED but it seems to predispose obese individuals to a higher BMI.	[63]
	rs1801260 3111T/C	C:54	241 women including 90 healthy controls, 60 patients with anorexia and 91 patients with bulimia.	Anorexia and Bulimia and BMI.	3111T/C SNP does not play a major role in anorexia and bulimia, but it seems to predispose eating disorders patients to more severe lifetime body weight loss.	[64]

	rs1801260 3111T/C	C:47	500 patients attending to a weight loss program based in the Mediterranean diet.	Weight loss.	Carriers of the minor C allele were more resistant to weight loss than TT individuals.	[51]
	rs1801260 3111T/C	C:47	1,500 patients attending to a weight loss program based in the Mediterranean diet.	Weight loss, implications of ghrelin and eveningness.	Minor C allele carriers had: (1) shorter sleep duration, (2) higher plasma ghrelin concentrations, (3) delayed breakfast time, (4) evening preference, and (5) less compliance with a Mediterranean Diet pattern.	[15]
	rs3749474 T/C rs1801260 T/C rs4864548 A/G rs1464490 T/C rs4580704 C/G	C:0.40 C: 0.37 G:0.40 C: 0.41 C:0.41 G:0.33	$N=1,100$ 540 men and 560 women who participated in the Genetics of Lipid Lowering Drugs and Diet Network (GOLDN) study.	CLOCK genetic variation and metabolic syndrome risk: modulation by monounsaturated fatty acids.	For SNP rs4580704, minor allele carriers had a 46% lower risk of hypertension than did noncarriers. By dichotomizing MUFA intake significant gene–diet interactions were identified associated with MetS.	[47]
	rs3749474 T/C rs1801260 T/C rs4864548 A/G rs1464490 T/C rs4580704 C/G	C:0.40 C: 0.37 G:0.40 C: 0.41 C:0.41 G:0.33	$N=1,100$; 540 men and 560 women who participated in the Genetics of Lipid Lowering Drugs and Diet Network (GOLDN) study.	Total energy intake.	Four of five CLOCK SNPs selected were significantly associated with total energy intake. For SNP rs3749474, the energy intake and total fat, protein, and carbohydrate intakes were higher in minor allele carriers than in non-carriers.	[48]
CLOCK/ SIRT1	Genotype combination SIRT1 (rs1467568) CLOCK (3111T/C, rs1801260)	Resistant genotype: *R*: 0.24	1,500 patients attending to a weight loss program based in the Mediterranean diet.	Weight loss.	Subjects carrying minor alleles at SIRT1 and CLOCK loci (R group) displayed a higher resistance to weight loss and a lower weekly weight loss rate as compared with homozygotes for both major alleles.	[56]

(continued)

Table 8.1 (continued)

Gene	SNP reference, genotype and Haplotype	Minor allele or Haplotype Frequency (%)	Population	Trait	Outcome	Reference
BMAL	59 SNPs from gene regions that showed preliminary evidence of associations with hypertension or Type 2 diabetes Haplotypes: rs7950226/ rs11022775 rs6486121/rs3789327/ rs969485.		1,304 individuals.	Hypertension and Type 2 diabetes.	After correcting for multiple testing (59 SNPs), none of the SNPs were significantly associated with hypertension or T2D. However, two haplotypes were associated with both pathologies.	[65]
PER2	rs2304672C>G and rs4663302C>T	G=0.06 T=0.34	500 individuals.	Abdominal obesity, psycho-behavioral factors, and attrition in the dietary treatment of obesity.	PER2 SNPs rs2304672C>G and rs4663302C>T were associated with abdominal obesity. Frequency of rs4663307 minor allele was greater in withdrawers. rs2304672 C>G minor allele carriers had a greater probability of dropping out, displaying extreme snacking, experiencing stress with dieting, eating when bored, and skipping breakfast than noncarriers.	[52]

BED binge eating disorders, *T2D* type 2 diabetes

CLOCK Interacts with Dietary Fat

A tantalizing finding from our research was the demonstration that these associations between the CLOCK gene and abdominal obesity or impaired glucose metabolism were only seen in individuals with dietary habits that included a high proportion of saturated fat (factory-made pastries, sausages, and so on) and a low proportion of monounsaturated fat (olive oil).

These results suggest that *CLOCK* polymorphisms interact with fatty acids to modulate MetS traits. Specifically, we identified significant gene–diet interactions associated to MetS at the CLOCK locus. By dichotomizing monounsaturated fatty acid (MUFA) intake, we found different effects across rs4580704 genotypes for glucose and insulin resistance. The protective effect of the minor allele on insulin sensitivity was only present when MUFA intake was high. Different effects were found across CLOCK 3111TNC genotypes for saturated fatty acid (SFA) intake. The deleterious effect of gene variants on waist circumference was only found with high SFA intakes [47]. These results suggest that dietary components (i.e., MUFA and SFA) are implicated in the relationship between alterations of circadian system and MetS.

CLOCK and Inflammation

As has been previously indicated, the quality and quantity of sleep could be associated with metabolic disruption and finally with obesity. Circadian clock genes are involved in sleep regulation [49, 50]. Therefore, the coordinated regulation of sleep or feeding behavior might be involved in the relationship between *CLOCK* SNPs and metabolic disorders. Along these lines, we have shown that the *CLOCK* locus was associated with plasma cytokine values, in particular with those that were involved in energy intake: MCP1, IL-6, and adiponectin [48]. Our data show strong associations between plasma IL-6 levels and the CLOCK rs3749474 SNP, which could be functionally related with changes in the *CLOCK* 3′-UTR structure (Fig. 8.3). Current data show that carriers of the minor allele, who reported high energy intake, also presented decreased plasma cytokine concentrations that could result, on the one hand, in a lower anorectic effect, and on the other, in decreased sleep

Moreover, we showed novel significant associations with individual MetS components such as waist, glucose metabolism-related variables and blood pressure. Carriers of the CGA (rs3749474/rs4580704/rs1801260 (3111T→C)) haplotype had lower BMI, waist circumference, blood pressure, and insulin resistance [46] (Fig. 8.3).

CLOCK and Weight Loss

Another study was performed in a Mediterranean population from south-east Spain, in a sample of 500 overweight/obese subjects, aged 20–65 years, who attended outpatient clinics specializing in obesity [51]. Consistent with a previous study, four

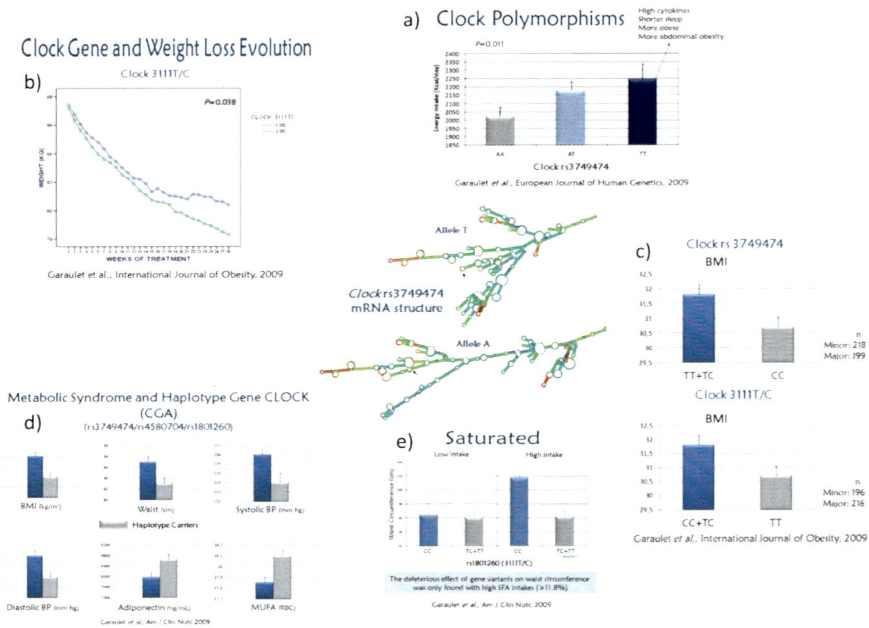

Fig. 8.3 *CLOCK* genetic variants and relationship with obesity and weight loss. (**a**) *CLOCK* rs 3749474and energy intake; (**b**) *CLOCK* 3111T/C and weight loss; (**c**) *CLOCK* SNPs and obesity; (**d**) Metabolic syndrome and haplotype *CLOCK* (CGA) (rs 3749474/rs4580704/rs1801260); (**e**) Nutrigenetics in *CLOCK*: interaction with saturated fat intake for obesity (adapted from [47, 48, 51]

out of five *CLOCK* SNPs selected were significantly associated with obesity variables. The genetic variation in the rs1801260 *CLOCK* was associated with obesity at baseline and also affected weight loss. Patients with the variant G allele lost significantly less weight compared with wild type patients.

Analysis of repeated measures showed that weight loss over time was significantly different between rs1801260 *CLOCK* variations. Carriers of the G allele displayed greater difficulty in losing weight than non-carriers. In this particular polymorphism, the frequency of short-time sleepers (≤6 h per day) was greater in minor G allele carriers than in non-carriers. *CLOCK* polymorphisms were also associated with significant differences in total plasma cholesterol at the completion of dietary treatment. It was concluded that the *CLOCK* rs1801260 SNP may predict the outcome of body weight reduction strategies based on low-energy diets. Furthermore, these results were replicated again in a bigger sample of 1,495 subjects, and this time we were also able to study different behaviors and metabolic variables in these subjects in order to elucidate which were the variables that were behind the association between this *CLOCK* SNP, weight loss and sleep (Fig. 8.4).

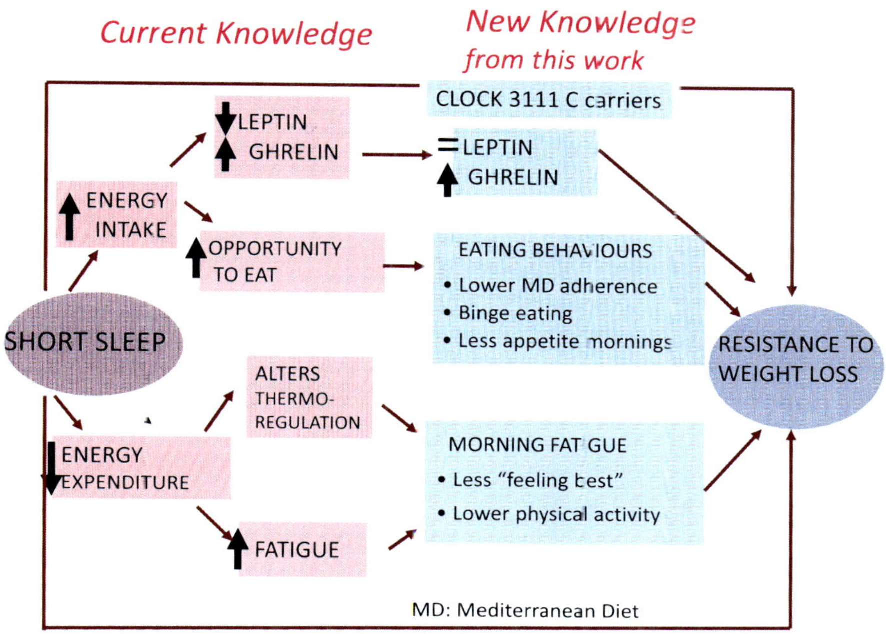

Fig. 8.4 Different behaviors and metabolic variables in CLOCK3111TC minor allele subjects (C carriers) behind the association between this SNP, weight loss, and sleep

PERIOD MetS and Obesogenic Behavior

As described above, various studies have shown that the presence of certain *PER2* gene polymorphisms such as (rs2304672C>G and rs4663302C>T) were associated with various psychological disturbances, particularly seasonal depression and bipolar disorder. This led us to consider whether a sample of overweight or obese patients would have emotional or psychological changes related to obesity and whether such changes would in turn be associated with *PER2* gene polymorphisms. Consistent with this idea, our results showed that people carrying the *PER2* gene variants were associated with abdominal obesity, and particularly *PER2* rs2304672 minor allele carriers (G) showed obesogenic behaviors, habits, and emotions, and greater rates of treatment discontinuation, nibbling, diet-induced stress, and food intake as an escape from boredom [52] (Fig. 8.5).

Another study performed in a population from Finland, also demonstrated the connection of *PER2* and MetS, in particular with high fasting blood glucose. *PER2* 10870 contributed to changes in glucose metabolism. *PER2*_10870 is an intronic mutation originally found by Spanagel et al. in 2005, when searching for the *PER2* SNPs modulating alcohol intake in mice. In this study performed by Englund et al. [53] *PER2*_10870 was associated with high fasting blood glucose. Another SNP in *PER2*

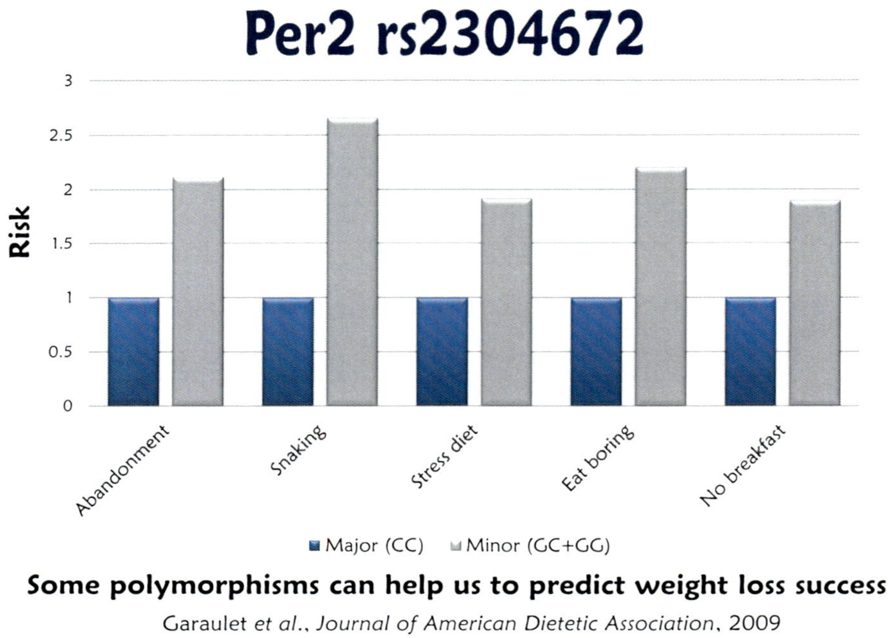

Fig. 8.5 *PER2* rs 2304672 and obesogenic behaviors (adapted from [51])

(*PER2* SNP rs934945) was also associated with waist circumference, and with MetS, although significance was lost after correcting for multiple comparisons [53].

BMAL1 and MetS

BMAL1 and *CLOCK* form a heterodimer and drive transcription from E-box elements found in the promoters of circadian-responsive genes. In addition to its roles in the control of circadian rhythms, *BMAL1* has been suggested to contribute to the metabolic regulation. Indeed, genome-wide profiling of *BMAL1* targets revealed their strict relationship with adipose tissue metabolism. Moreover, SNP analysis revealed that BMAL1 is associated with type II diabetes and hypertension [54]. However, replications are needed in different populations to determine the extension of this associations and the specific functional SNPs influencing MetS risk.

Gene × Gene Interactions in the Clock

Different gene by gene interactions between clock genes have been related to MetS, obesity or weight loss. This is the case of a significant interaction between the *5-HTTLPR* variant of the serotonin gene and the haplotype rs1554483–rs4864548 in *CLOCK* that was associated with diastolic and systolic blood pressure, arterial hypertension, plasma triglycerides and different MetS components [55]. In all these cases, the higher values were observed in rotating shift workers homozygous for the *SLC6A*4 S allele and carrying the haplotype composed by the *CLOCK* rs1554483 G and rs4864548 A variants. These data suggest a potential interaction (epistatic effect) of serotonin transporter and *CLOCK* gene variation on the genetic susceptibility to develop MS by rotating shift workers.

Other example of the gene × gene interaction is *SIRT1* and *CLOCK* 3111T>C combined genotype which is associated with evening preference and weight loss resistance in a behavioral therapy treatment for obesity. Our group has demonstrated that variants at both *SIRT1* and *CLOCK* have an additive effect on resistance to weight loss that could be related to the chronotype of the subject, higher plasma levels of ghrelin and less adherence to Mediterranean diet patterns [56].

Epigenetics and the Clock System

As indicated earlier, the world of epigenetics is revolutionizing genetics. Epigenetic research shows that we are not predetermined by our genome. What we eat, how much we sleep, if we exercise or even how we use our mind may change our epigenome, and our fate. Moreover, these changes are not restrained to us but can pass down to our children or even to our grant children. In other words, epigenetics does not change the DNA but decides how much or whether some genes are expressed in different cells in our bodies.

The molecular basis of epigenetics is complex. It involves modifications of the activation of certain genes, but not the core DNA structure. One way that gene expression is regulated is by the remodeling of chromatin (the complex of DNA and histones). Chromatin proteins associated with DNA may be activated or silenced. Histones can change how tightly or loosely the DNA wraps around them by modifying their amino acids. If the amino acids that are in the chain are changed, the shape of the histone sphere might be modified.

A second way of chromatin remodeling is the addition of methyl groups to the DNA, mostly at CpG sites, which are regions of DNA where a cytosine nucleotide occurs next to a guanine nucleotide. "CpG" is shorthand for "—C—phosphate—G—", that is, cytosine and guanine separated by only one phosphate. Methylation converts cytosine to 5-methylcytosine. Some areas of the genome are methylated more heavily than others, and highly methylated areas tend to be less transcriptionally active.

Epigenetic control can be exerted through a variety of mechanisms, including not only DNA methylation but also microRNA-mediated metabolic pathways, histone variants and histone Post Translational Modifications or PTMs. Different studies point out to the association between epigenetics changes and several illnesses, for example cancer, in which methylation of CpG sites within the promoters of genes can lead to the silencing of tumor suppressor genes. In contrast, the hypomethylation of CpG sites has been associated with the over-expression of oncogenes within cancer cells.

In 2011, it was demonstrated for the first time that the methylation of mRNA had a critical role in human energy homeostasis. Obesity associated FTO gene was shown to be able to demethylate N6-methyladenosine in RNA. This opened the related field of RNA epigenetics and its relation to obesity.

Epigenetics and Circadian Rhythms

The connection between epigenetics and the clock machinery came out with the study of Crosio et al. [57], who demonstrated that chromatin remodeling was involved in circadian gene expression. These authors showed that a pulse of light, when applied to mice during the subjective night, induced histone phosphorylation in the SCN. This was an early effect which implicated an induction of *Per1* translation. Subsequently, it has been indicated that histone modifications at clock controlled genes promoters occur in a circadian manner.

It has been hypothesized that because the number of transcripts that oscillates in a circadian manner is high, there must be a widespread program of dynamic changes in chromatin remodeling that accompany circadian gene expression. This has been described as the "circadian epigenome" and probably includes cycles of chromatin transitions that allow a highly dynamic chromatin structure to be temporally permissive to transcription.

Histones can be modified at more than 30 sites within the N-terminal tails. There are several modifications in the histones such as acetylation and phosphorylation, among others. The finding that *CLOCK* has an intrinsic Histone Acetile Transferase (HAT) activity reveals one link between epigenetic control and the circadian clock. *CLOCK* may acetylate histones, particularly the lysine residues in the histones 3 and 4. Interestingly, *CLOCK* can also acetylate non-histone substrates. This is the case of *BMAL1* which is acetylated by *CLOCK* in a lysine residue, an event which is crucial for the circadian transcriptional program. Other substrate susceptible to be acetylated by *CLOCK* is the glucocorticoid receptor, whose function is regulated by this process. This acetylation activity by *CLOCK* has been demonstrated to be essential for circadian expression.

Another important player in the circadian epigenome, is sirtuin, particularly *SIRT1* and *SIRT6*. Sirtuins possess deacetylase activity and are implicated in the induction of gene silencing. Particularly *SIRT1* physically interacts with *CLOCK* and it has been defined as a circadian enzymatic rheostat [2], because it controls by different mechanisms the balance of acetylation and chromatin remodeling by the circadian clock.

Despite this conceptual knowledge, data about the connection between this circadian epigenome and obesity are still scarce. Nevertheless, recently our group has demonstrated an association between the methylation status of CpG sites located in clock genes (CLOCK, BMAL1 and PER2) with obesity, Metabolic Syndrome and weight loss [58].

The Role of MicroRNAs in the Clock System

MicroRNAs are small (approximately 22 nucleotides), single-stranded, noncoding RNA species that act as potent gene silencers and are relevant players in different physiological and pathophysiological processes. They can be divided in three types depending o their characteristics: one type is the "*intergenic microRNA*" which are those microRNAs which are coded within introns and exons of noncoding RNA; the second type are the "*introgenic microRNA*" which are coded from introns of protein encoding genes, and the third type are the "*polycistronic microRNAs*" long RNA that contains multiple microRNAs in a cluster. The gene silencing function of microRNAs occurs via a reduction in mRNA translation efficiency or mRNA stability.

Considering that microRNAs appear to be involved in the regulation of >60% of human genes, it is not surprise that they have been identified as major players in the regulation of the circadian rhythm [59]. Therefore, they are emerging as new potential therapeutic targets for disorders of the circadian clock based on the following features:

(a) MicroRNAs *present in the SCN* can control the core clock molecular feedback by posttranscriptional-mediated control mechanisms [60].
(b) MicroRNAs *display circadian rhythmicity, in some cases regulated by light*, such as Arabidopsis miR167, miR168, miR171 and miR398, whereas in others, like the fly miR263b, the regulation is directly *coordinated by the circadian clock* [61].
(c) Some microRNAs have been related to *circadian alterations. In vivo* antisense silencing studies demonstrated that miR219 shortens the circadian period and that miR132 negatively regulates light-dependent resetting of the clock [60]. Both miR132 and miR219 affect *per1* expression, and thus influence the core circadian transcriptional loop [60, 62]. However, neither miR132 nor miR219 appears to directly target *per1*; thus, the precise mechanisms by which these microRNAs affect the clock are not known.
(d) In addition to rhythmic microRNA expression in the SCN clock, *peripheral oscillators* also exert circadian control over microRNA expression. For example, some microRNAs such as the miR192/194 cluster are highly expressed in the liver and kidneys (two oscillating organs). Others such as miR16, miR20a and miR141, oscillate in the intestine, and several microRNAs have been found to exhibit diurnal oscillations in the mouse retina [59, 62].

Given the importance of microRNAs it has been proposed that SNPs mapping with them may have functional consequences resulting in phenotypic variation. This is the case of a polymorphism in pre-miR182 which exhibits diurnal rhythmicity in the retina and has been significantly associated with major depression [63].

This promising preliminary evidence suggests that following a better understanding of microRNA biology; these may become a new therapeutic tool in the fight against obesity.

Conclusion

A number of genetic variants connect circadian rhythmicity and obesity. This new knowledge should be considered on the prevention and treatment of obesity. Moreover, it is important to expand the current nutrigenetic knowledge to guide the health professionals on the delivery of more tailored advice based on genetic variants.

Summary

1. Thanks to the drive infused by the completion of the HGP, a number of technologies and research approaches have evolved. Nowadays, new functional genomics areas to accomplish the goals started by the HGP. Nutrigenetics, nutrigenomics, and epigenetics are some examples.
2. The living matter has its own time, i.e., the "biological time." In this sense, we could assume that "a Chronome exists into the Genome":"
3. In the last years the study of genetics and its relationship to chronobiology has been focusing in the chronotype and different psychological illnesses associated to these chronotype such as depression, anxiety or bipolar disorder.
4. Some examples of genetic variants associated to psychological traits are *CLOCK* 3111T/C associated with evenningness, and different personality traits; *PERIOD2* (*PER2* rs2304672) polymorphism which moderates circadian-relevant reward circuitry; *CRY1* (rs8192440) related to psychological treatment effectiveness; *CRY2* associated with winter depression and *SIRT1* rs10997875 a good candidate gene for the pathophysiology for mood disorders.
5. Other candidate SNPs connected to obesity could be those associated to sleep disorders. Some examples are serotonin receptors, prepro-orexin or IL-6 SNPs which associate with obstructive sleep apnea syndrome. Others are SNPs residing in ROR1 and PLCB1which associated with insomnia.
6. One particular example that could link sleep duration and obesity is *ABCC9;* the leukocyte antigen *(HLA) DQB1*0602; PER3* (G-320T, C -319A) and *NPAS2* rs6725296 or *BMA1* rs6290035 TT which had greater loadings on the metabolic factor (weight and appetite) of the global seasonality score.
7. Sleep disorders or short sleep duration, are both associated to several polymorphisms connected to obesity. In this regard, one of the best studied *CLOCK* SNPs (3111TC) has been significantly associated to short sleep duration, eveningness, several psychological traits, weight loss and obesity.

8. *Epigenetics* is the study of heritable changes in gene expression or cellular phenotype caused by mechanisms other than changes in the underlying DNA sequence. *CLOCK* and *SIRT* have both epigenetics effect in acetylation and deacetylation of histones, respectively. MicroRNAs may become novel therapeutic targets for disorders in the circadian clock. The knowledge achieved in the circadian epigenome could give us new answers to the connections among genetics, circadian rhythmicity, and obesity.

References

1. Pietro Cugini (1993) Chronobiology: principles and methods - medical semeiology and methodology copyright: Annali Istituto Superiore di Sanità 29: 483–500
2. Bellet MM, Sassone-Corsi P (2010) Mammalian circadian clock and metabolism—the epigenetic link. J Cell Sci 123(Pt 22):3837–3848
3. Eckel-Mahan KL, Patel VR, Mohney RP, Vignola KS, Baldi P, Sassone-Corsi P (2012) Coordination of the transcriptome and metabolome by the circadian clock. Proc Natl Acad Sci U S A 109(14):5541–5546
4. Dallmann R, Viola AU, Tarokh L, Cajochen C, Brown SA (2012) The human circadian metabolome. Proc Natl Acad Sci U S A 109(7):2625–2629
5. Soria V, Martínez-Amorós E, Escaramís G, Valero J, Pérez-Egea R, García C, Gutiérrez-Zotes A, Puigdemont D, Bayés M, Crespo JM, Martorell L, Vilella E, Labad A, Vallejo J, Pérez V, Menchón JM, Estivill X, Gratacòs M, Urretavizcaya M (2010) Differential association of circadian genes with mood disorders: CRY1 and NPAS2 are associated with unipolar major depression and CLOCK and VIP with bipolar disorder. Neuropsychopharmacology 35(6):1279–1289
6. McClung CA (2007) Circadian genes, rhythms and the biology of mood disorders. Pharmacol Ther 114(2):222–232
7. Sprouse J, Braselton J, Reynolds L (2006) Fluoxetine modulates the circadian biological clock via phase advances of suprachiasmatic nucleus neuronal firing. Biol Psychiatry 60:896–899
8. Abe M, Herzog ED, Block GD (2000) Lithium lengthens the circadian period of individual suprachiasmatic nucleus neurons. Neuroreport 11:3261–3264
9. Xu X, Breen G, Chen CK, Huang YS, Wu YY, Asherson P (2010) Association study between a polymorphism at the 3′-untranslated region of CLOCK gene and attention deficit hyperactivity disorder. Behav Brain Funct 6:48
10. Serretti A, Benedetti F, Mandelli L, Lorenzi C, Pirovano A, Colombo C, Smeraldi E (2003) Genetic dissection of psychopathological symptoms insomnia in mood disorders and CLOCK gene polymorphism. Am J Med Genet B Neuropsychiatr Genet 121B:35–38
11. Takao T, Tachikawa H, Kawanishi Y, Mizukami K, Asada T (2007) CLOCK gene T3111C polymorphism is associated with Japanese schizophrenics: a preliminary study. Eur Neuropsychopharmacol 17:273–276
12. Benedetti F, Dallaspezia S, Fulgosi MC, Lorenzi C, Serretti A, Barbini B, Colombo C, Smeraldi E (2007) Actimetric evidence that CLOCK 3111 T/C SNP influences sleep and activity patterns in patients affected by bipolar depression. Am J Med Genet B Neuropsychiatr Genet 144B:631–635
13. Benedetti F, Radaelli D, Bernasconi A, Dallaspezia S, Falini A, Scotti G, Lorenzi C, Colombo C, Smeraldi E (2008) Clock genes beyond the clock: CLOCK genotype biases neural correlates of moral valence decision in depressed patients. Genes Brain Behav 7:20–25
14. Kishi T, Kitajima T, Ikeda M, Yamanouchi Y, Kinoshita Y, Kawashima K, Okochi T, Okumura T, Tsunoka T, Inada T, Ozaki N, Iwata N (2009) Association study of clock gene (CLOCK)

and schizophrenia and mood disorders in the Japanese population. Eur Arch Psychiatry Clin Neurosci 259:293–297
15. Garaulet M, Sánchez-Moreno C, Smith CE, Lee YC, Nicolás F, Ordovás JM (2011) Ghrelin, sleep reduction and evening preference: relationships to CLOCK 3111 T/C SNP and weight loss. PLoS One 6(2):e17435
16. Lavebratt C, Sjöholm LK, Soronen P, Paunio T, Vawter MP, Bunney WE, Adolfsson R, Forsell Y, Wu JC, Kelsoe JR, Partonen T, Schalling M (2010) CRY2 is associated with depression. PLoS One 5(2):e9407
17. McCarthy MJ, Nievergelt CM, Shekhtman T, Kripke DF, Welsh DK, Kelsoe JR (2011) Functional genetic variation in the Rev-Erbα pathway and lithium response in the treatment of bipolar disorder. Genes Brain Behav 10(8):852–861
18. Kishi T, Yoshimura R, Kitajima T, Okochi T, Okumura T, Tsunoka T, Yamanouchi Y, Kinoshita Y, Kawashima K, Fukuo Y, Naitoh H, Umene-Nakano W, Inada T, Nakamura J, Ozaki N, Iwata N (2010) SIRT1 gene is associated with major depressive disorder in the Japanese population. J Affect Disord 126(1–2):167–173
19. Utge SJ, Soronen P, Loukola A, Kronholm E, Ollila HM, Pirkola S, Porkka-Heiskanen T, Partonen T, Paunio T (2010) Systematic analysis of circadian genes in a population-based sample reveals association of TIMELESS with depression and sleep disturbance. PLoS One 5(2):e9259
20. Dauvilliers Y, Tafti M (2008) The genetic basis of sleep disorders. Curr Pharm Des 14(32):3386–3395
21. Ban HJ, Kim SC, Seo J, Kang HB, Choi JK (2011) Genetic and metabolic characterization of insomnia. PLoS One 6(4):e18455
22. Zhang X, Liu RY, Lei Z, Zhu Y, Huang JA, Jiang X, Liu Z, Liu X, Peng X, Hu H, Zhang HT (2009) Genetic variants in interleukin-6 modified risk of obstructive sleep apnea syndrome. Int J Mol Med 23(4):485–493
23. Larkin EK, Patel SR, Zhu X, Tracy RP, Jenny NS, Reiner AP, Walston J, Redline S (2010) Study of the relationship between the interleukin-6 gene and obstructive sleep apnea. Clin Transl Sci 3(6):337–339
24. Miaskowski C, Dodd M, Lee K, West C, Paul SM, Cooper BA, Wara W, Swift PS, Dunn LB, Aouizerat BE (2010) Preliminary evidence of an association between a functional interleukin-6 polymorphism and fatigue and sleep disturbance in oncology patients and their family caregivers. J Pain Symptom Manage 40(4):531–544
25. Greenberg H, Cohen RI (2012) Nocturnal asthma. Curr Opin Pulm Med 18(1):57–62, Review
26. Vardeny O, Peppard PE, Finn LA, Faraco JH, Mignot E, Hla KM (2011) β2 adrenergic receptor polymorphisms and nocturnal blood pressure dipping status in the Wisconsin Sleep Cohort Study. J Am Soc Hypertens 5(2):114–122
27. Chen W, Ye J, Han D, Yin G, Wang B, Zhang Y (2012) Association of prepro-orexin polymorphism with obstructive sleep apnea/hypopnea syndrome. Am J Otolaryngol 33(1):31–36
28. Park HJ, Park JK, Kim SK, Cho AR, Kim JW, Yim SV, Chung JH (2011) Association of polymorphism in the promoter of the melatonin receptor 1A gene with schizophrenia and with insomnia symptoms in schizophrenia patients. J Mol Neurosci 45(2):304–308
29. Olsson L, Pettersen E, Ahlbom A, Carlsson S, Midthjell K, Grill V (2011) No effect by the common gene variant rs10830963 of the melatonin receptor 1B on the association between sleep disturbances and type 2 diabetes: results from the Nord-Trøndelag Health Study. Diabetologia 54(6):1375–1378
30. Shimada M, Miyagawa T, Kawashima M, Tanaka S, Honda Y, Honda M, Tokunaga K (2010) An approach based on a genome-wide association study reveals candidate loci for narcolepsy. Hum Genet 128(4):433–441
31. He Y, Jones CR, Fujiki N, Xu Y, Guo B, Holder JL Jr, Rossner MJ, Nishino S, Fu YH (2009) The transcriptional repressor DEC2 regulates sleep length in mammals. Science 325(5942):866–870

32. Science News. The ABCC9 of sleep: a genetic factor regulates how long we sleep. Science Daily (Nov. 24, 2011)
33. Goel N, Dinges DF (2011) Behavioral and genetic markers of sleepiness. J Clin Sleep Med 7(5 Suppl):S19–S21, Review
34. Archer SN, Carpen JD, Gibson M, Lim GH, Johnston JD, Skene DJ, von Schantz M (2010) Polymorphism in the PER3 promoter associates with diurnal preference and delayed sleep phase disorder. Sleep 33(5):695–701
35. Vandewalle G, Archer SN, Wuillaume C, Balteau E, Degueldre C, Luxen A, Maquet P, Dijk DJ (2009) Functional magnetic resonance imaging-assessed brain responses during an executive task depend on interaction of sleep homeostasis, circadian phase, and PER3 genotype. J Neurosci 29(25):7948–7956
36. Kovanen L, Saarikoski ST, Aromaa A, Lönnqvist J, Partonen T (2010) ARNTL (BMAL1) and NPAS2 gene variants contribute to fertility and seasonality. PLoS One 5(4):e10007
37. Science News. Shorter sleep durations may increase genetic risks for obesity. Science Daily (June 15, 2011)
38. Prats-Puig A, Grau-Cabrera P, Riera-Pérez E, Cortés-Marina R, Fortea E, Soriano-Rodríguez P, de Zegher F, Ibánez L, Bassols J, López-Bermejo A (2012) Variations in the obesity genes FTO, TMEM18 and NRXN3 influence the vulnerability of children to weight gain induced by short sleep duration. Int J Obes (Lond). Doi:10.1038/ijo.2012.27. [Epub ahead of print]
39. Laposky AD, Shelton J, Bass J et al (2006) Altered sleep regulation in leptin-deficient mice. Am J Physiol Regul Integr Comp Physiol 290:R894–R903
40. Laposky AD, Bradley MA, Williams DL et al (2008) Sleep-wake regulation is alteredin leptin resistant (db/db) genetically obese and diabetic mice. Am J Physiol Regul Integr Comp Physiol 295:R2059–R2066
41. Turek FW, Joshu C, Kohsaka A, Lin E, Ivanova G, McDearmon E, Laposky A, Losee-Olson S, Easton A, Jensen DR, Eckel RH, Takahashi JS, Bass J (2005) Obesity and metabolic syndrome in circadian clock mutant mice. Science 308:1043–1045
42. Rudic RD, McNamara P, Curtis AM, Boston RC, Panda S, Hogenesch JB, Fitzgerald GA (2004) BMAL1 and CLOCK, two essential components of the circadian clock, are involved in glucose homeostasis. PLoS Biol 2:e377
43. Kennaway DJ, Owens JA, Voultsios A, Boden MJ, Varcoe TJ (2007) Metabolic homeostasis in mice with disrupted clock gene expression in peripheral tissues. Am J Physiol Regul Integr Comp Physiol 293:R1528–R1537
44. Shimba S, Ogawa T, Hitosugi S, Ichihashi Y, Nakadaira Y, Kobayashi M, Tezuka M, Kosuge Y, Ishige K, Ito Y, Komiyama K, Okamatsu-Ogura Y, Kimura K, Saito M (2011) Deficient of a clock gene, brain and muscle Arnt-like protein-1 (BMAL1), induces dyslipidemia and ectopic fat formation. PLoS One 6(9):e25231
45. Sookoian S, Gemma C, Gianotti TF, Burgueño A, Castaño G, Pirola CJ (2008) Genetic variants of Clock transcription factor are associated with individual susceptibility to obesity. Am J Clin Nutr 87(6):1606–1615
46. Scott EM, Carter AM, Grant PJ (2008) Association between polymorphisms in the Clock gene, obesity and the metabolic syndrome in man. Int J Obes (Lond) 32(4):658–662
47. Garaulet M, Lee YC, Shen J, Parnell LD, Arnett DK, Tsai MY, Lai CQ, Ordovas JM (2009) CLOCK genetic variation and metabolic syndrome risk: modulation by monounsaturated fatty acids. Am J Clin Nutr 90(6):1466–1475
48. Garaulet M, Lee YC, Shen J, Parnell LD, Arnett DK, Tsai MY, Lai CQ, Ordovas JM (2010) Genetic variants in human CLOCK associate with total energy intake and cytokine sleep factors in overweight subjects (GOLDN population). Eur J Hum Genet 18(3):364–369
49. Katzenberg D, Young T, Finn L, Lin L, King DP, Takahashi JS, Mignot E (1998) A CLOCK polymorphism associated with human diurnal preference. Sleep 21(6):569–576
50. Xu Y, Toh KL, Jones CR, Shin JY, Fu YH, Ptácek LJ (2007) Modeling of a human circadian mutation yields insights into clock regulation by PER2. Cell 128(1):59–70

51. Garaulet M, Corbalán MD, Madrid JA, Morales E, Baraza JC, Lee YC, Ordovas JM (2010) CLOCK gene is implicated in weight reduction in obese patients participating in a dietary programme based on the Mediterranean diet. Int J Obes (Lond) 34(3):516–523
52. Garaulet M, Corbalán-Tutau MD, Madrid JA, Baraza JC, Parnell LD, Lee YC, Ordovas JM (2010) PERIOD2 variants are associated with abdominal obesity, psycho-behavioral factors, and attrition in the dietary treatment of obesity. J Am Diet Assoc 110(6):917–921
53. Englund A, Kovanen L, Saarikoski ST, Haukka J, Reunanen A, Aromaa A, Lönnqvist J, Partonen T (2009) NPAS2 and PER2 are linked to risk factors of the metabolic syndrome. J Circadian Rhythms 7:5
54. Hatanaka F, Matsubara C, Myung J, Yoritaka T, Kamimura N, Tsutsumi S, Kanai A, Suzuki Y, Sassone-Corsi P, Aburatani H, Sugano S, Takumi T (2010) Genome-wide profiling of the core clock protein BMAL1 targets reveals a strict relationship with metabolism. Mol Cell Biol 30(24):5636–5648
55. Sookoian S, Gianotti TF, Burgueño A, Pirola CJ (2010) Gene–gene interaction between serotonin transporter (SLC6A4) and CLOCK modulates the risk of metabolic syndrome in rotating shiftworkers. Chronobiol Int 27(6):1202–1218
56. Garaulet M, Esteban Tardido A, Lee YC, Smith CE, Parnell LD, Ordovás JM (2012) SIRT1 and CLOCK 3111T>C combined genotype is associated with evening preference and weight loss resistance in a behavioral therapy treatment for obesity. Int J Obes (Lond). Doi:10.1038/ijo.2011.270. [Epub ahead of print]
57. Crosio C, Cermakian N, Allis CD, Sassone-Corsi P (2000) Light induces chromatin modification in cells of the mammalian circadian clock. Nat Neurosci 3(12):1241–1247
58. Fermín I. Milagro, Purificación Gómez-Abellán, Javier Campión, J. Alfredo Martínez, Jose M. Ordovás and Marta Garaulet (2012) CLOCK, PER2 and BMAL1 DNA Methylation: Association with Obesity and Metabolic Syndrome Characteristics and Monounsaturated Fat Intake, Chronobiology International, Early Online: 1–15, Copyright © Informa Healthcare USA, Inc. ISSN 0742-0528 print/1525-6073 online DOI: 10.3109/07420528.2012.719967
59. Hansen KF, Sakamoto K, Obrietan K (2011) MicroRNAs: a potential interface between the circadian clock and human health. Genome Med 3(2):10
60. Cheng HY, Papp JW, Varlamova O, Dziema H, Russell B, Curfman JP, Nakazawa T, Shimizu K, Okamura H, Impey S, Obrietan K (2007) MicroRNA modulation of circadian-clock period and entrainment. Neuron 54:813–829
61. Yang M, Lee JE, Padgett RW, Edery I (2008) Circadian regulation of a limited set of conserved microRNAs in Drosophila. BMC Genomics 9:83
62. Alvarez-Saavedra M, Antoun G, Yanagiya A, Oliva-Hernandez R, Cornejo-Palma D, Perez-Iratxeta C, Sonenberg N, Cheng HY (2011) MiRNA-132 orchestrates chromatin remodeling and translational control of the circadian clock. Hum Mol Genet 20:731–751
63. Saus E, Soria V, Escaramís G, Vivarelli F, Crespo JM, Kagerbauer B, Menchón JM, Urretavizcaya M, Gratacòs M, Estivill X (2010) Genetic variants and abnormal processing of pre-miR-182, a circadian clock modulator, in major depression patients with late insomnia. Hum Mol Genet 19:4017–4025
64. Monteleone P, Tortorella A, Docimo L, Maldonato MN, Canestrelli B, De Luca L, Maj M (2008) Investigation of 3111T/C polymorphism of the CLOCK gene in obese individuals with or without binge eating disorder: association with higher body mass index. Neurosci Lett 435:30–33
65. Tortorella A, Monteleone P, Martiadis V, Perris F, Maj M (2007) The 3111T/C polymorphism of the CLOCK gene confers a predisposition to a lifetime lower body weight in patients with anorexia nervosa and bulimia nervosa: a preliminary study. Am J Med Genet B Neuropsychiatr Genet 144B:992–995
66. Woon PY, Kaisaki PJ, Bragança J, Bihoreau MT, Levy JC, Farrall M, Gauguier D (2007) Aryl hydrocarbon receptor nuclear translocator-like (BMAL1) is associated with susceptibility to hypertension and type 2 diabetes. Proc Natl Acad Sci U S A 104:14412–14417

Chapter 9
Chronobiology and Metabolic Syndrome: From Genes to Systems Biology

Silvia Sookoian and Carlos J. Pirola

Abstract The major function of the circadian system is the internal cycling of physiologic and metabolic events. In the last years, there has been an exponential increase in our understanding of the role of clock-related genes in Metabolic-syndrome (MetS)-related phenotypes. Nevertheless, our understanding about how the components of the circadian system interact each other to modulate the metabolism and cardiovascular system remains a major challenge.

Systems biology introduces a new concept for revealing the pathogenesis of human disorders and suggests the presence of common physiologic processes and molecular networks influencing the risk of a disease. In this review, we use systems biology approaches to integrate genomic, molecular, and physiological data to decipher putative circadian rhythmic pathways suspected to play a role in the etiology of the metabolic syndrome (MetS)-associated phenotypes with a main focus in obesity and as other authors are discussing diverse related topics, we discuss the findings of our own group.

S. Sookoian, M.D., Ph.D. (✉)
Instituto de Investigaciones Médicas A. Lanari,
Av. Combatiente de Malvinas 3150, (1427) Buenos Aires, Argentina

Department of Clinical and Molecular Hepatology, Institute of Medical Research A Lanari-IDIM, University of Buenos Aires-National-Council of Scientific and Technological Research (CONICET), Ciudad Autónoma de Buenos Aires, Argentina
e-mail: sookoian.silvia@lanari.fmed.uba.ar

C.J. Pirola, Ph.D. (✉)
Department of Molecular Genetics and Biology of Complex Diseases, Institute of Medical Research A Lanari-IDIM, University of Buenos Aires-National-Council of Scientific and Technological Research (CONICET), Ciudad Autónoma de Buenos Aires, Argentina
e-mail: pirola.carlos@lanari.fmed.uba.ar

M. Garaulet and J.M. Ordovás (eds.), *Chronobiology and Obesity*,
DOI 10.1007/978-1-4614-5082-5_9, © Springer Science+Business Media New York 2013

Abbreviations

MetS	Metabolic syndrome
IR	Insulin resistance
T2D	Type-2 diabetes
GWAS	Genome-wide association scan
CLOCK	Circadian locomotor output cycles kaput protein
5-HT	Serotonin
5-HTT	Serotonin transporter
5HIAA	5-Hydroxyindolacetic acid
PPARG1A	Peroxisome proliferator-activated receptor gamma cofactor 1 alpha gene

Introduction

The Impact of the Circadian System in the Metabolic Syndrome-Associated Phenotypes

Metabolic syndrome (MetS) is associated with several metabolic disturbances, including insulin resistance (IR) in several tissues. Indeed, IR is considered as the main link among all the clinical disorders clustered in MetS, namely type-2 Diabetes (T2D), dyslipidemias, central obesity, arterial hypertension, prothrombotic and proinflammatory states, ovarian polycystosis, and nonalcoholic fatty liver disease (NAFLD). From the perspective of clinical importance, MetS has two prominent features: Its worldwide prevalence that is dramatically increasing and its strong association with cardiovascular disease, as initially described by Reaven G [1].

The pathobiology of the MetS results from a complex interplay between genes and environment. Actually, environmental factors, such as decreased physical activity, increased nutrient availability, and overnutrition, play an important role in the development of metabolic disturbances associated with IR, and are also largely recognized as responsible for the modern epidemic of MetS-related phenotypes.

The genetic component of each individual component of the MetS has been largely investigated and both genome-wide and candidate gene association studies have identified several loci that influence the susceptibility of all the clustering traits [2, 3].

Nevertheless, the gene variants identified so far by genome-wide association scans (GWAS) do not explain by themselves the pathophysiology of the MetS, and they account for a modest effect on the disease. There has been a loci that was not initially identified by any of the GWAS but showed an interesting biological plausibility and a significant effect on some intermediate phenotypes associated with MetS, such as obesity. This locus is related to the circadian system, and is actually the master gene circadian locomotor output cycles kaput protein (*CLOCK*).

Actually, the importance of maintaining the internal homeostasis of the circadian systems and its impact on human MetS was recently revealed by Turek et al. by an animal model in which homozygous *Clock* mutant mice had a greatly attenuated diurnal feeding rhythm, were hyperphagic and obese, and developed hyperleptinemia, hyperlipidemia, hepatic steatosis, hyperglycemia, and hypoinsulinemia [4]. This finding is biologically plausible as the major function of the circadian system is the internal cycling of physiologic and metabolic events [5].

Given the above mentioned evidence and the results of further emerging studies, which show that altering circadian rhythmicity results in pathophysiological changes resembling MetS and fat accumulation, we immediately explored the role of gene variants and derived haplotypes of the *CLOCK* transcription factor in obesity and related quantitative metabolic traits in a human study [6]. Hence, we recruited in a case–control design, 715 lean and 391 overweight or obese unrelated adult subjects, and investigated six tag single-nucleotide polymorphisms (SNPs) with a minor allele frequency >10% (rs1554483 C/G; rs11932595 A/G; rs4580704 C/G; rs6843722 A/C; rs6850524 C/G, and rs4864548 A/G) encompassing 117 kb of chromosome 4 and representing 115 polymorphic sites. The results of our study suggested that *CLOCK* polymorphism and related haplotypes are critically involved in the genetic susceptibility to obesity as carrying the haplotype of rs1554483G and rs4864548A was associated with 1.8-fold increase for being overweight/obese [6]. These findings were replicated in other populations around the word [7–9]. In addition, although the association of the *CLOCK* variants with other classical components of the MetS such as arterial hypertension did not persist after adjusting for overweight/obesity, we did find a significant association with non alcoholic fatty liver disease (NAFLD) in a hospital-based study [10]. In fact, the *CLOCK* variants, rs11932595 and rs6843722 showed significant associations with NAFLD (empiric $P=0.0449$ and 0.023, respectively), and a significant association was also observed between clinical or histologic spectrum of the disease and rs1554483 (empiric $P=0.0399$), rs6843722 (empiric $P=0.0229$) and rs6850524 (empiric $P=0.00899$), and between fibrosis score and rs1554483 (empiric $P=0.02697$), rs6843722 (empiric $P=0.01898$) and rs4864548 (empiric $P=0.02697$) suggesting a potential role of the *CLOCK* polymorphisms and their haplotypes in the susceptibility to NAFLD and disease severity [10]. A more comprehensive review of the genetic basis of the MetS was reported elsewhere [3].

In summary, while in the last years there has been an exponential increase in our understanding of the role of clock-related genes in MetS-related phenotypes [11], to know how the components of the circadian system interact with each other to modulate the metabolism and cardiovascular system remains a major challenge.

A Short Overview About Circadian System–Gene Components

The circadian system of mammals is composed of a hierarchy of oscillators that function at the cellular, tissue, and systems levels (Fig. 9.1) [12]. Component of the

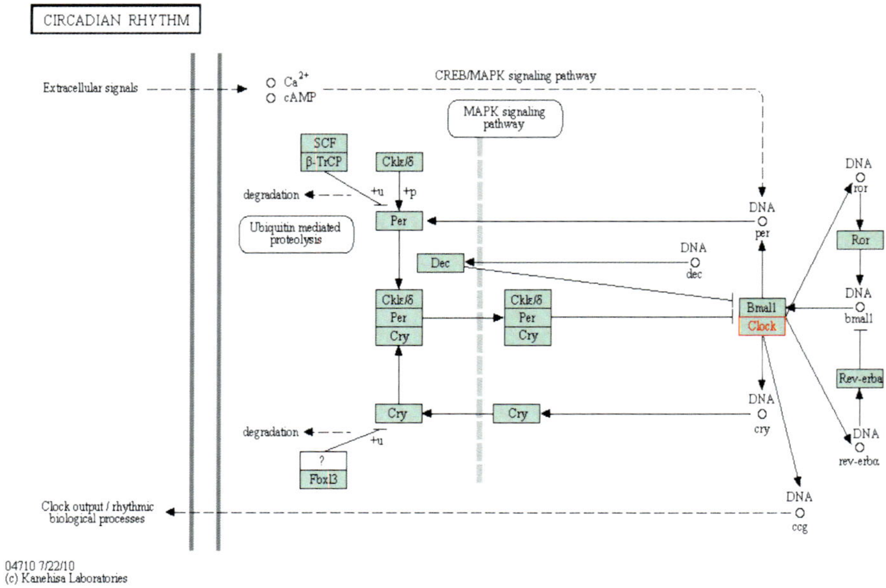

Fig. 9.1 Circadian rhythm system component of mammal—*Homo sapiens* (human) according to the KeGG (Kyoto Encyclopedia of Genes and genomes) map 04710

circadian clock oscillator includes the CRY proteins, CLOCK or NPAS2, ARNTL or ARNTL2, CSNK1D and/or CSNK1E, TIMELESS and the PER proteins. Table 9.1 summarizes the main features of the transcription/translation feedback loops of the mammalian circadian clock. As this picture is an oversimplification, here, we decide to explore the association of obesity and circadian rhythm-related genes using systems biology.

The Use of a Systems Biology Approach to Better Understand the Association of Circadian Rhythm-Related Genes with Obesity and MetS

Systems biology introduces a new concept for revealing the pathogenesis of human disorders and suggests the presence of common physiologic processes and molecular networks influencing the risk of a disease. Rather than compartmentalizing individual risk factors (e.g., IR, blood pressure, body mass index, or lipid concentrations) and treating them as if they were separate and independent, systems biology examines their interactions. Here we show a model of this concept to explain the impact of the circadian system on one of the most important features of the MetS, obesity, by different systems-biology approaches, which are mostly based on gene enrichment analysis and protein–protein interaction networks.

Table 9.1 Short overview about the main function and features of the genes related with the circadian system

Gene symbol	Gene name	Main function and features
CLOCK	Circadian locomotor output cycles kaput protein	• Protein involved in the generation of rhythmic pattern of behaviors or activities, e.g., circadian rhythm which is a metabolic or behavioral rhythm within a cycle of 24 h. • ARNTL/2-CLOCK heterodimers activate E-box element transcription of a number of proteins of the circadian clock. • Activates transcription of PER1 and PER2. • Has intrinsic histone acetyltransferase activity and this enzymatic function contributes to chromatin-remodeling events implicated in circadian control of gene expression. • CLOCK gene, homolog to murine Clock, regulator of circadian rhythms, with two major transcripts 8 and 10 kb, predominantly expressed in the suprachiasmatic nuclei and cerebellum. • Expressed in all peripheral tissues, including spleen, thymus, prostate, testis, ovary, small intestine, colon, leukocytes, heart, brain, placenta, lung, liver, skeletal muscle, kidney, adipose tissue and pancreas.
CRY1	Cryptochrome 1 (photolyase-like)	• Blue light-dependent regulator of the circadian feedback loop. Inhibits CLOCK NPAS2-ARNTL E box-mediated transcription. • Acts, in conjunction with CRY2, in maintaining period length and circadian rhythmicity. • Capable of translocating circadian clock core proteins such as PER proteins to the nucleus. • Expression is regulated by light and circadian rhythms. Peak expression in the suprachiasmatic nucleus (SCN) and eye at the day/night transition (CT12). Levels decrease with ARNTL-CLOCK inhibition as part of the autoregulatory feedback loop. • Belongs to the DNA photolyase class-1 family.
CRY2	Cryptochrome 2 (photolyase-like)	• Blue light-dependent regulator of the circadian feedback loop • Translocated to the nucleus through interaction with other Clock proteins such as PER2 or ARNTL • Belongs to the DNA photolyase class-1 family. • Regulate the functional activity of circadian transcriptional complex at the posttranslational level • Phosphorylation on Ser-266 by MAPK is important for the inhibition of CLOCK-ARNTL-mediated transcriptional activity. Phosphorylation by CSKNE requires interaction with PER1 or PER2. • Expressed in all tissues examined including fetal brain, fibroblasts, heart, brain, placenta, lung, liver, skeletal muscle, kidney, pancreas, spleen, thymus, prostate, testis, ovary, small intestine, colon, and leukocytes. Highest levels in heart and skeletal muscle.

(continued)

Table 9.1 (continued)

Gene symbol	Gene name	Main function and features
PER1/PER2	Period homolog 1 (Drosophila) and period homolog 2 (Drosophila)	• Component of the circadian clock mechanism which is essential for generating circadian rhythms. • Negative element in the circadian transcriptional loop. Influences clock function by interacting with other circadian regulatory proteins and transporting them to the nucleus. • Negatively regulates CLOCK NPAS2-BMAL1 BMAL2-induced transactivation. • PER1: Widely expressed. Found in heart, brain, placenta, lung, liver, skeletal muscle, pancreas, kidney, spleen, thymus, prostate, testis, ovary, and small intestine. Highest level in skeletal muscle. Low level in kidney. • PER2: Widely expressed. Found in heart, brain, placenta, lung, liver, skeletal muscle, kidney, adipose tissue and pancreas. High levels in skeletal muscle and pancreas. Low level in lung.
ARNTL	Aryl hydrocarbon receptor nuclear translocator-like (alias BMAL)	• The protein encoded by this gene is a basic helix-loop-helix protein that forms a heterodimer with CLOCK. • This complex binds an E-box upstream of the PER1 gene, activating this gene and possibly other circadian rhythm-associated genes. • master regulator of circadian rhythm, also playing important roles in the regulation of adipose differentiation and lipogenesis in mature adipocytes.
CSNK1D	Casein kinase 1, delta	• This gene is a member of the casein kinase I (CKI) gene family whose members have been implicated in the control of cytoplasmic and nuclear processes, including DNA replication and repair. • Phosphorylates connexin-43/GJA1, MAP1A, SNAPIN, MAPT/TAU, TOP2A, DCK, HIF1A, EIF6, p53/TP53, DVL2, DVL3, ESR1, AIB1/NCOA3, DNMT1, PKD2, YAP1, PER1, and PER2. • Central component of the circadian clock. May act as a negative regulator of circadian rhythmicity by phosphorylating PER1 and PER2, leading to retain PER1 in the cytoplasm.
TIMELESS	*Timeless* homolog (Drosophila)	• Required for normal progression of S-phase. • Involved in the circadian rhythm autoregulatory loop. • Negatively regulates CLOCK-NPAS2/BMAL1-induced transactivation of PER1 possibly via translocation of PER1 into the nucleus. • Expressed in all tissues examined including brain, heart, lung, liver, skeletal muscle, kidney, placenta, pancreas, spleen, thymus, and testis. Highest levels of expression in placenta, pancreas, thymus and testis. • Belongs to the timeless family.

For example, we used a bioinformatic resource named "Platform for Exploration of Significant Concepts Associated to co-Ocurrences Relationships" (Pescador, cbdm.mdc-berlin/tools/pescador) [13].

PESCADOR uses LAITOR-Literature Assistant for Identification of Terms co-Occurrences and Relationships as text-mining engine to extract sentences with co-occurring bioentities (genes and proteins) from the text of the PubMed abstracts requested [13]. Thus, PESCADOR allows selecting gene–protein co-occurrence pairs based on their relatedness to biological concepts and therefore brings together under a common perspective protein interactions that have not been studied under the same research focus [13].

Thus, we used the keywords, "obesity AND genetics AND circadian" retrieving 200 references. The associated genes under this search are described in Table 9.2. Interestingly, more than 100 genes associated with the regulation of food intake and energy expenditure, glucose and lipid metabolism, and of course circadian rhythmicity, as displayed in Fig. 9.2, participate in a highly interconnected interactome. Interestingly, the hierarchical order of the genes shows that *CLOCK* is the most important one, followed by other components of the cell circadian clock mechanism (Fig. 9.3). The notion of "interactome" includes the complete list of physical interactions mediated by all proteins of an organism [14]. Indeed, biological questions are increasingly addressed in the framework of such complex molecular networks (Tables 9.3 and 9.4).

Based on the hypothesis that common physiologic processes and molecular networks influence the risk of obesity, we proposed another systems biology approach: a gene enrichment analysis evaluated by the bioinformatic resource *ToppGene Suite* (http://toppgene.cchmc.org) [15]. A similar concept was recently applied for finding new candidate genes for the MetS [2], rendering new loci whose associations with MetS components were finally replicated in independent studies, i.e., *HNF4A* with type 2 diabetes in more than 49,000 individuals by a meta-analysis [16], and *IGF1R* with arterial hypertension in an Argentinean population [17]. Importantly, by using the ToppFun (Transcriptome, ontology, phenotype, proteome, and pharmacome annotations based gene list functional enrichment analysis, a tool of the Toppgene suite), is possible to obtain a general picture of the biological process these genes are belonging to with a significant P value $< 1 \times 10^{-6}$. In this sense, it is important to note that in addition to the obvious pathways already mentioned, an important number of loci, i.e., 25%, belong to biological process involved in regulation of the transcriptional activity of polymerase II probably as nuclear factors "per se" or through their activity on epigenetic marks, i.e., histone deacetylases (HDACs, SIRT1, etc.). This topic is further developed below.

Some other genes are involved in the response to xenobiotics (according to ToppFun, P value $< 1 \times 10^{-6}$), and then it is worth of noting that not only are some clinically relevant drugs but natural compound mentioned. In this regards, it is particularly interesting that several serotonin transporter inhibitors, such as sibutramine or fenfluramine and serotonin itself are included.

Table 9.2 List of genes retrieved by the Pescador tool (available at http://cbdm.mdc-berlin.de/tools/pescador/)

Gene symbol	Gene name
ACACB	Acetyl-CoA carboxylase beta
ACSS2	Acyl-CoA synthetase short-chain family member 2
ADIPOQ	Adiponectin, C1Q and collagen domain containing
ADIPOR1	Adiponectin receptor 1
ADIPOR2	adiponectin receptor 2
AGRP	Agouti related protein homolog (mouse)
ARNTL	Aryl hydrocarbon receptor nuclear translocator-like
ATP2A2	ATPase, Ca2+ transporting, cardiac muscle, slow twitch 2
BDNF	Brain-derived neurotrophic factor
BTG2	BTG family, member 2
C2	Complement component 2
CCRN4L	CCR4 carbon catabolite repression 4-like (*S. cerevisiae*)
CD36	CD36 molecule (thrombospondin receptor)
CLOCK	Clock homolog (mouse)
COL4A5	Collagen, type IV, alpha 5
CREM	cAMP responsive element modulator
CRY1	Cryptochrome 1 (photolyase-like)
CRY2	Cryptochrome 2 (photolyase-like)
CYP19A1	Cytochrome P450, family 19, subfamily A, polypeptide 1
DBP	D site of albumin promoter (albumin D-box) binding protein
ETV3	Ets variant 3
FAS	Fas (TNF receptor superfamily, member 6)
GAPDH	Glyceraldehyde-3-phosphate dehydrogenase
GCG	Glucagon
GCGR	Glucagon receptor
GCK	Glucokinase (hexokinase 4)
GHRH	Growth hormone releasing hormone
GHRL	Ghrelin/obestatin prepropeptide
GLB1	Galactosidase, beta 1
GLP1R	Glucagon-like peptide 1 receptor
GLP2R	Glucagon-like peptide 2 receptor
GPR50	G protein-coupled receptor 50
GPT2	Glutamic pyruvate transaminase (alanine aminotransferase) 2
H6PD	Hexose-6-phosphate dehydrogenase (glucose 1-dehydrogenase)
HCRT	Hypocretin (orexin) neuropeptide precursor
HDAC1	Histone deacetylase 1
HDAC3	Histone deacetylase 3
HR	Hairless homolog (mouse)
ID1	Inhibitor of DNA binding 1, dominant negative helix–loop–helix protein
ID2	Inhibitor of DNA binding 2, dominant negative helix–loop–helix protein
INS	Insulin
LEP	Leptin
LEPR	Leptin receptor
LTF	Lactotransferrin
MC3R	Melanocortin 3 receptor

(continued)

Table 9.2 (continued)

Gene symbol	Gene name
MC4R	Melanocortin 4 receptor
MECP2	Methyl CpG binding protein 2 (Rett syndrome)
MT1E	Metallothionein 1E
MT2A	Metallothionein 2A
MTNR1A	Melatonin receptor 1A
MTNR1B	Melatonin receptor 1B
NAMPT	Nicotinamide phosphoribosyltransferase
NCOR1	Nuclear receptor corepressor 1
NFIL3	Nuclear Factor, interleukin 3 regulated
NOVA2	Neuro-oncological ventral antigen 2
NPAS2	Neuronal PAS domain protein 2
NPY	Neuropeptide Y
NR1I2	Nuclear receptor subfamily 1, group I, member 2
NR1I3	Nuclear receptor subfamily 1, group I, member 3
NR3C1	Nuclear receptor subfamily 3, group C, member 1 (glucocorticoid receptor)
NUCB2	Nucleobindin 2
O3FAR1	Omega-3 fatty acid receptor 1
PAIP2	Poly(A) binding protein interacting protein 2
PCK2	Phosphoenolpyruvate carboxykinase 2 (mitochondrial)
PDK4	Pyruvate dehydrogenase kinase, isozyme 4
PER1	Period homolog 1 (Drosophila)
PER2	Period homolog 2 (Drosophila)
PER3	Period homolog 3 (Drosophila)
POMC	Proopiomelanocortin
PPA1	Pyrophosphatase (inorganic) 1
PPARA	Peroxisome proliferator-activated receptor alpha
PPARG	Peroxisome proliferator-activated receptor gamma
PRKAA1	Protein kinase, AMP-activated, alpha 1 catalytic subunit
PRL	Prolactin
PROK2	Prokineticin 2
PROKR2	Prokineticin receptor 2
PYY	Peptide YY
SDC3	Syndecan 3
SDC3 s	Yndecan 3
SERPINE1	Serpin peptidase inhibitor, clade E (nexin, plasminogen activator inhibitor type 1), member 1
SLC2A4	Solute carrier family 2 (facilitated glucose transporter), member 4
SLC5A1	Solute carrier family 5 (sodium/glucose cotransporter), member 1
SLC5A1	Solute carrier family 5 (sodium/glucose cotransporter), member 1
SLC9A3	Solute carrier family 9 (sodium/hydrogen exchanger), member 3
STAT3	Signal transducer and activator of transcription 3 (acute-phase response factor)
UBE4A	Ubiquitination factor E4A
UCP1	Uncoupling protein 1 (mitochondrial, proton carrier)
UCP3	Uncoupling protein 3 (mitochondrial, proton carrier)
VGF	VGF nerve growth factor inducible
YARS	Tyrosyl-tRNA synthetase

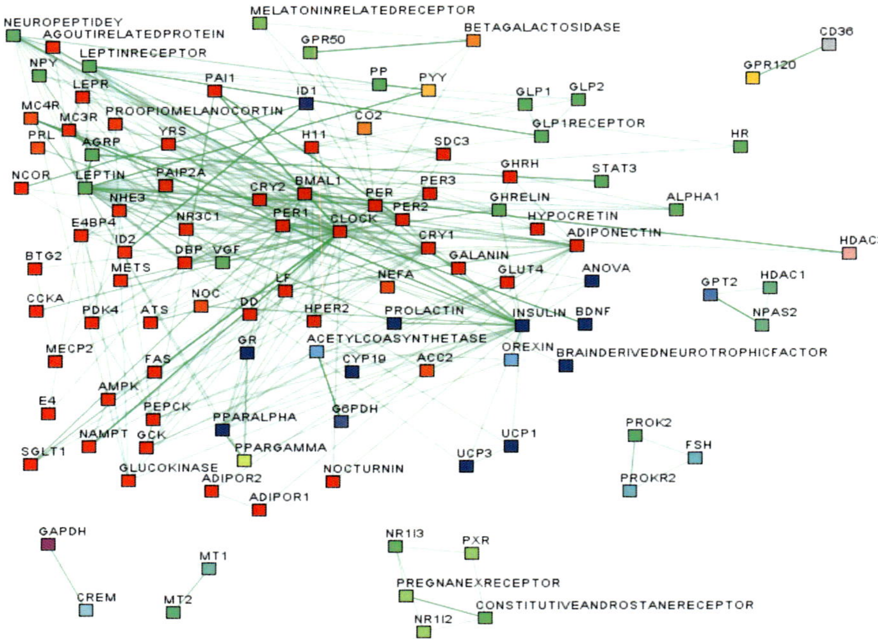

Fig. 9.2 Interactome of genes listed in Table 9.2

The Connection Between the Circadian System, Serotonin and Obesity

As an example of the application of systems biology to understand the pathobiology of human diseases, we explain in detail the results of a highly predicted protein by the above mentioned strategy: the serotonin transporter and its association with obesity.

Serotonin transporter (5-HTT) is involved in mood and eating disturbances and encoded by the gene *SLC6A4*, the promoter shows functional insertion/deletion alleles: long (L) and short (S). Because individuals who are carriers for the short version are known to be at risk for higher levels of anxiety, we hypothesized that this variant may be associated with overweight [18, 19]. We collected data and blood samples from 172 adolescents out of a cross-sectional, population-based study of 934 high school students and to replicate the findings, we also included 119 outpatients from the Nutrition and Diabetes Section of the Children's County Hospital [18]. We found that the S allele was associated with overweight (BMI > 85th percentile), being a risk factor for overweight independently of sex, age, and hypertension [odds ratio (OR): 1.85; 95% confidence interval (CI): 1.13, 3.05; $P < 0.02$] [18]. Additionally, in the outpatient study, compared with the homozygous LL subjects, S allele carriers showed a higher BMI z-score (1.47 ± 1.09 vs. 0.51 ± 1.4; $P < 0.002$)

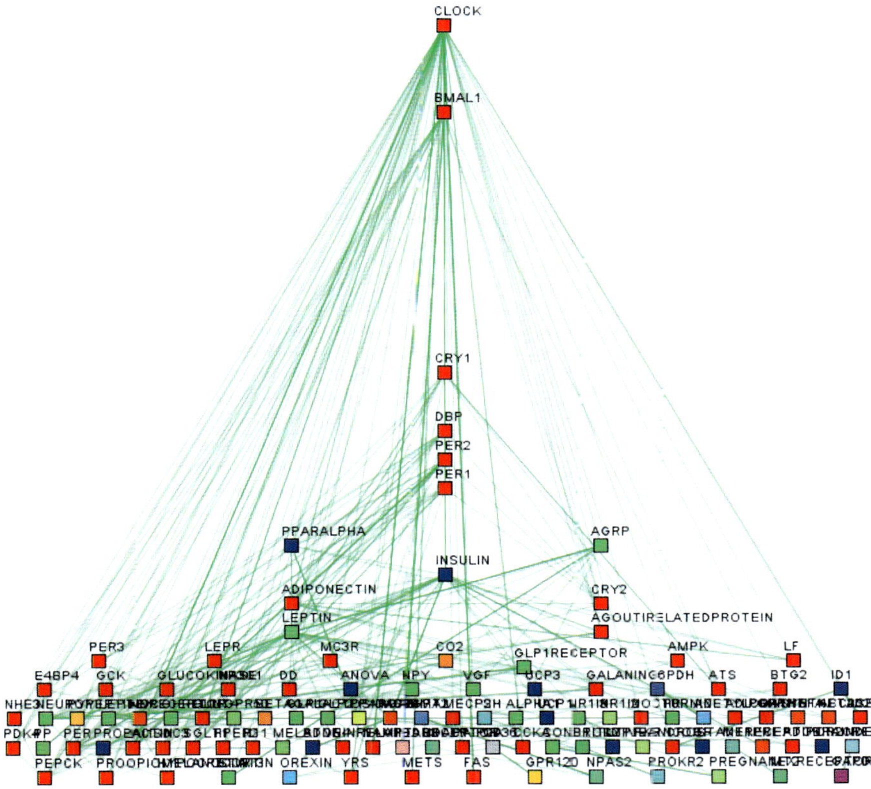

Fig. 9.3 Hierarchical interactome of genes listed in Table 9.2

and were more frequent in overweight children [18]. Furthermore, we further evaluated whether the S/L variant of the *SLC6A4* gene is associated with BMI as a continuous trait and also with obesity in a large sample of adult men of European ancestry included in a cross-sectional, population-based study (1,329 unrelated subjects, aged 34.6 ± 0.3 years) [19]. We observed statistically significant differences across genotypic groups (LL: 25.4 ± 0.2, LS: 26.0 ± 0.[illegible] and SS: 26.7 ± 0.2, $P < 0.0002$). In addition, association tests showed that the 5-HTTLPR-genotype distribution was significantly different between 692 lean (BMI ≤ 25 kg/m^2) and 637 overweight/obese (BMI ≥ 27 kg/m^2) individuals, and we found a 1.36 odds ratio (OR) (95% CI 1.01–1.85) for obesity in SS carriers in comparison with LL carriers, $P = 0.026$ [19]. Our findings indicate that 5-HTTLPR polymorphism may be linked with BMI and also with obesity and/or overweight in adolescent and adult populations reinforcing the role of the serotonin transporter as a risk factor for the obesity phenotype and suggest potential new avenues for its pharmacological treatment.

In fact, because serotonin (5-HT) is a neurotransmitter associated with circadian rhythm regulation, we explored a possible relation among 5-HT, its metabolite,

Table 9.3 Biological process to which listed genes are involved with as classified by Gene Ontology (GO) terms

ID of the GO term	Name of the GO term	Hit count in query list	Hit count in genome
GO:0007623	Circadian rhythm	21	85
GO:0048511	Rhythmic process	25	201
GO:0009725	Response to hormone stimulus	29	644
GO:0009719	Response to endogenous stimulus	29	714
GO:0010033	Response to organic substance	35	1,206
GO:0032870	Cellular response to hormone stimulus	19	345
GO:0043434	Response to peptide hormone stimulus	19	353
GO:0071495	Cellular response to endogenous stimulus	19	363
GO:0007631	Feeding behavior	12	94
GO:0006091	Generation of precursor metabolites and energy	19	443
GO:0070887	Cellular response to chemical stimulus	24	819
GO:0009605	Response to external stimulus	27	1,123
GO:0071310	Cellular response to organic substance	19	492
GO:0019318	Hexose metabolic process	15	261
GO:0031327	Negative regulation of cellular biosynthetic process	23	796
GO:0006006	Glucose metabolic process	14	217
GO:0009755	Hormone-mediated signaling pathway	11	102
GO:0009890	Negative regulation of biosynthetic process	23	810
GO:0032787	Monocarboxylic acid metabolic process	18	445
GO:0048878	Chemical homeostasis	22	740
GO:0055114	Oxidation–reduction process	24	935
GO:0032094	Response to food	7	22
GO:0007610	Behavior	18	477
GO:0031667	Response to nutrient levels	15	297
GO:0005996	Monosaccharide metabolic process	15	299
GO:0032102	Negative regulation of response to external stimulus	10	87
GO:0048545	Response to steroid hormone stimulus	15	303
GO:0032098	Regulation of appetite	6	13
GO:0032095	Regulation of response to food	6	13
GO:0042592	Homeostatic process	25	1,086
GO:0006366	Transcription from RNA polymerase II promoter	26	1,206
GO:0006357	Regulation of transcription from RNA polymerase II promoter	24	1,015
GO:0009892	Negative regulation of metabolic process	25	1,112
GO:0009991	Response to extracellular stimulus	15	325
GO:0010817	Regulation of hormone levels	16	386
GO:0031324	Negative regulation of cellular metabolic process	24	1,023
GO:0048585	Negative regulation of response to stimulus	12	174
GO:0010558	Negative regulation of macromolecule biosynthetic process	21	773
GO:0043436	Oxoacid metabolic process	22	863
GO:0019752	Carboxylic acid metabolic process	22	863
GO:0042593	Glucose homeostasis	9	73
GO:0033500	Carbohydrate homeostasis	9	73

(continued)

Table 9.3 (continued)

ID of the GO term	Name of the GO term	Hit count in query list	Hit count in genome
GO:0006082	Organic acid metabolic process	22	879
GO:0042180	Cellular ketone metabolic process	22	880
GO:0040014	Regulation of multicellular organism growth	9	78
GO:0032868	Response to insulin stimulus	13	249
GO:0032107	Regulation of response to nutrient levels	6	18
GO:0032104	Regulation of response to extracellular stimulus	6	18
GO:0006066	Alcohol metabolic process	18	580
GO:0046879	Hormone secretion	12	207

The Gene Ontology project is a major bioinformatics initiative with the aim of standardizing the representation of gene and gene product attributes across species and databases. The project provides a controlled vocabulary of terms for describing gene product characteristics and gene product annotation data from GO Consortium members, as well as tools to access and process this data (http://www.geneontology.org/)

5-hydroxyindolacetic acid (5HIAA), and the functional polymorphism of the serotonin transporter gene (SLC6A4) promoter with rotating shift work. We decided to explore the impact of serotonin on this clinical model, as rotating shift work disrupts the synchronous relationship between the body's internal clock and the environment, causing deleterious effects not only on hormonal levels but also on neurotransmitters and activity–rest cycles. Thus, we performed a cross sectional study, including 683 men, and 437 day workers were compared with 246 rotating shift workers, and we found that platelet 5-HT content differed significantly ($P=0.002$) between day workers (41.28 ± 1.99 pg/mg) and rotating shift workers (37.91 ± -4.16 pg/mg); 5-HIAA content was also significantly ($P=0.00004$) higher in day workers (11.40 ± 0.82 pg/mg) than in rotating shift workers (9.33 ± 1.02 pg/mg). When we looked for differences in *SLC6A4* promoter, we found a significant ($P=0.016$) difference in genotype distribution between day workers LL: 126 (28.8%), LS: 202 (46.2%), and SS: 109 (24.9%), and rotating shift workers LL: 47 (19.1%), LS: 124 (50.4%), and SS: 75 (30.5%). Interestingly, when we divided the subjects between workers with less and more than 60 month rotating shift-work exposure, the difference in *SLC6A4* genotypes frequency was only significant in the group with > or =60 months ($P=0.011$). In addition, there was a significantly lower content of platelet 5-HIAA in S allele carriers in comparison with the other genotypes (SS: 9.2 ± 1.0 pg/mg vs. SL/LL: 11.0 ± 0.8 pg/mg, $P<0.02$). As a conclusion, platelet 5-HT and 5-HIAA contents were significantly lower in rotating shift workers than day workers, and there was a significant association between the S variant of SLC6A4 promoter and shift work. These findings may be important for targeting effective therapeutic strategies to ameliorate the associated comorbidities and behavioral problems in rotating shift workers, because, as it is well-known, rotating shift workers present many of the features of the MetS including low-grade inflammation and elevated circulating cytokines [20–22].

Table 9.4 Interacting xenobiotics with the listed genes

ID	Name	Source	Hit count in query list	Hit count in genome
CID000004829	Pioglitazone	Stitch	25	326
CID000000896	Melatonin	Stitch	26	443
CID000005753	Corticosterone	Stitch	24	381
CID000003003	Dexamethasone	Stitch	38	1,343
D005632	Fructose	CTD	16	125
CID000000861	Triiodothyronine	Stitch	23	409
CID000004599	Orlistat	Stitch	14	86
CID000003413	Forskolin	Stitch	32	1,014
D003907	Dexamethasone	CTD	26	624
CID000000274	Cyclic AMP	Stitch	30	907
MESH:D009765/ D005632-M	Obesity affected by fructose	CTD Marker	8	11
CID000005591	Troglitazone	Stitch	24	526
CID000004091	Metformin	Stitch	18	235
CID000005300	Streptozotocin	Stitch	24	540
CID000001106	Stearoyl-coenzyme A	Stitch	14	122
CID000102191	2-Deoxyglucose	Stitch	18	272
CID000000869	Malonyl-COA	Stitch	14	126
CID000005210	Sibutramine	Stitch	10	40
D005947	Glucose	CTD	14	143
CID000002900	Cycloheximide	Stitch	27	906
CID000005244	Dehydroepiandrosterone sulfate	Stitch	17	277
CID000005694	Wy-14,643	Stitch	17	287
CID000060303	Englitazone	Stitch	8	22
CID000077999	Rosiglitazone	Stitch	22	605
CID000145068	Nitric oxide	Stitch	28	1,076
CID000003339	Fenofibrate	Stitch	16	276

MESH:D009765/ D004041-M	Obesity affected by Dietary Fats	CTD Marker	8	27
CID000031275	1,4-Dioxane	Stitch	16	278
CID000003640	Cortisol	Stitch	20	504
CID000005408	Testosterone	Stitch	21	578
D010126	Ozone	CTD	25	882
CID000004920	Progesterone	Stitch	28	1,171
CID000004585	Olanzapine	Stitch	12	129
C023036	Perfluorooctanoic acid	CTD	20	543
CID000005202	Serotonin	Stitch	20	554
CID000005098	GI262570	Stitch	11	104
D005227	Fatty Acids	CTD	8	34
CID000000304	Cholesterol	Stitch	21	637
CID000003337	Fenfluramine	Stitch	13	178
CID000004174	Metyrapone	Stitch	13	179
CID000000681	Dopamine	Stitch	22	715
CID000002826	Cocaine	Stitch	20	571
CID000002833	Colchicine	Stitch	17	383
C055122	Orlistat	CTD	9	56
CID000003285	Estrogen	Stitch	28	1,252
C009250	Sevoflurane	CTD	12	146
CID000003779	Isoproterenol	Stitch	21	669
CID000000702	EtOH	Stitch	23	837
CID000000076	Dehydroepiandrosterone	Stitch	15	292
D019324	Beta-naphthoflavone	CTD	20	608
CID000036523	Synthetic LH-RH	Stitch	14	246
CID000000200	5-Aminoimidazole-4-carboxamide ribotide	Stitch	12	163

(continued)

Table 9.4 (continued)

ID	Name	Source	Hit count in query list	Hit count in genome
CID000017513	AICA	Stitch	11	126
CID000000985	Palmitate	Stitch	16	367
CID000000085	Carnitine	Stitch	14	265
CID005310993	Acipimox	Stitch	9	69
D002211	Capsaicin	CTD	11	133
CID000001424	13-HODE	Stitch	11	136

The Circadian System Is Linked to a Master Regulator of Energy Metabolism: The Transcriptional Coactivator, Peroxisome Proliferative Activated Receptor Gamma Coactivator 1 Alpha (PPARGC1A)

We used another bioinformatic tool, Genemania (GeneMANIA.org [23]), to extend the previous list with functionally similar genes that the program identifies using available genomics and proteomics data reporting weights that indicate the predictive value of each selected data set for the query. As depicted in Fig. 9.4, among the new loci predicted is the peroxisome proliferator-activated receptor gamma cofactor 1 alpha gene (*PPARG1A*, also known as *PGC1A*). The transcriptional coactivator *PPARG1A* coordinates the regulation of genes involved in energy metabolism by controlling transcriptional programs of mitochondrial biogenesis, adaptive thermogenesis, and fatty-acid betaoxidation. In fact, its tissue specificity pattern of expression is mainly located in the heart, skeletal muscle, liver, and kidney [24].

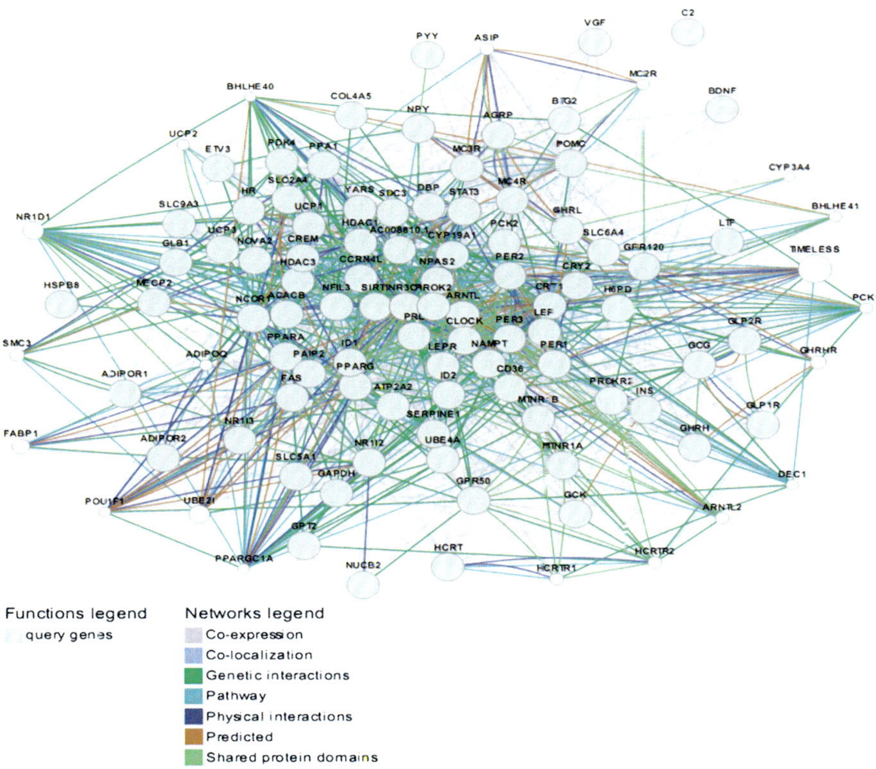

Fig. 9.4 Genemania predictive network using interactions as depicted in the legend. New loci are rendered (*open circles*). The *arrow* indicated the PPARG cofactor 1 alpha (PPARGC1A, also known as PGC1A)

Interestingly, the protein encoded by this gene is involved in controlling blood pressure, regulating cellular cholesterol homoeostasis, and development of obesity, and altered signaling of *PPARGC1A* contributes to glucose intolerance, IR, and T2D [25, 26]. It is worthy to mention that we observed that methylation levels of the *PPARGC1A* promoter in the liver tissue of NAFLD patients are correlated with HOMA-IR and plasma fasting insulin levels [27]. Interestingly, we also observed that liver abundance of *PPARGC1A* mRNA is inversely correlated with the methylation levels of *PPARGC1A* promoter CpGs, and also with the status of peripheral IR, suggesting that methylation of, at least, three explored CpG sites in the gene promoter efficiently repressed its transcriptional activity [27]. Finally, we were able to show that mitochondrial biogenesis is reduced in the liver of NAFLD patients and is associated with peripheral IR and *PPARGC1A* promoter methylation status [27].

Expert Opinion: The Problem of the Missing Heritability

The effect of isolated genes on the disease susceptibility is in general moderate, being odds ratios (ORs) smaller than 2, at best. Even considering the sum of the effect of several risk variants combined (i.e., for seven hypothetical variants with a minor allele frequency of 0.5 and a OR of 1.5, the overall composite effect is smaller than 10 and the probability of finding individuals of such haplotypes would be less than 5% in the total population) the total variance of the phenotype explained is commonly less than 5%. A clear example can be found in the study by Li et al. [28] who showed that individuals who carried >16 risk alleles for obesity had higher BMI than those who carried <7 risk alleles but by only 1.53 BMI units and all SNPs add only 3% to the predictive value of obesity in addition to age and sex.

Then, we should considered the possibility that risk variants are not acting independently but rather by a synergistic effect, a phenomenon known as epistasis. Owing to the difficulties of the study design, few authors have reported the role of epistasis on the risk of MetS-associated phenotypes, for instance, hypertension [29], myocardial infarction and coronary artery disease [30], cholesterol levels [31], triglyceridemia [32]. In this vein, we have shown that *CLOCK* and serotonin transporter (*SLC6A4*) variants interacting with environmental factors, such as the rotating shift work have an strong effect on the overall MetS development and/or its isolated components, such as blood pressure or plasma triglycerides [33].

Recent advances in epigenomic approaches have placed the epigenetic gene regulation as a key factor in the pathogenesis of many complex disorders, mostly cancer but also other complex diseases, such as MetS, as mentioned above. In particular, epigenetic modifications can explain the mechanisms involved in the gene–environment interaction, the sexual dimorphism observed in some phenotypes, and the role of developmental programming. The most attractive aspect of the hypothesis of the impact of epigenetic changes in the etiology of the MetS-associated phenotypes is given by the nature of the epigenetic regulation, which is dynamic and subjected to either external or internal influences. Hence, while epigenetic marks can be propagated during cell division resulting in a permanent modification of the phenotype, they are also plausible of therapeutic modifications.

The most studied mechanism of epigenetic modifications influencing a MetS-related trait is DNA methylation. For instance, DNA methylation of the *PPARGC1A* promoter in pancreatic islets from type 2 diabetic patients was associated with alterations in the insulin secretion [34]. In this vein, there is increasing evidence that prenatal environment can modify the epigenetic regulation of specific genes. We recently reported a positive correlation between maternal BMI and *PPARGC1A* promoter methylation in umbilical cord of their offspring's, suggesting a potential role of promoter *PPARGC1A* methylation in the metabolic programming of the fetus [35]. In addition, we explored whether DNA methylation of the mitochondrial transcription factor A (*TFAM*) promoter is associated with insulin resistance in adolescents with features of MS and observed a potential role of epigenetic modifications in this transcription factor in association with HOMA-IR and fasting plasma insulin levels [36]. Interestingly, because *PPARGC1A* and *TFAM* are, both, regulators of mitochondrial biogenesis, DNA methylation at their promoters may be one of cause of the mitochondrial DNA decrease we have observed in small and large for gestational age newborns as well as adolescents with insulin resistance [37, 38].

A Practical Information: How to Apply This New Concepts in the Clinical Practice

The systems biology approach is designed to analyze and integrate genomic, transcriptomic, and/or proteomic data to infer from genetic signals related pathways of disease. MetS is a complex constellation of diseases, thus it is possible to infer that common physiologic processes and molecular networks influence the risk of each intermediate phenotype. Accumulating evidence indicates that circadian desynchronization and/or alterations in circadian clock gene function have both a strong impact on the maintaining of metabolic homeostasis and cardiovascular function.

The components of the circadian system can be regarded as a network of complex interactions working either synergistically or in an integrated system that modulate the susceptibility of MetS associated phenotypes.

The most paradigmatic example in clinical practice of disruption of circadian rhythmicity is the observed in workers under rotating shift work schedules, which represents an important risk factor for the development of MetS [20, 21]. This point is particularly relevant because in industrialized nations and modern societies, as many as 20% of workers are rotating shift workers [39].

Although the epidemiological and clinical studies have provided a well-documented association among shift work schedule and MetS and cardiovascular disease, the molecular mechanisms underlying these phenomena are not yet well understood. Systems biology may shed light on novel putative pathobiological pathways associated with the disease and may also suggest novel therapeutic approaches. For instance, the interplay of serotonin, light, and the clock circadian system and their impact on metabolic functions may provide Health professionals with new insight into the etiology and the treatment of MetS-related phenotypes in rotating shift workers.

Summary Points

- The major function of the circadian system is the internal cycling of physiologic and metabolic events to adapt cell metabolism to external cues such as sleep–awake rhythm, light–night cycle, and food availability.
- Altering the circadian rhythmicity results in pathophysiological changes inducing MetS and fat accumulation.
- Polymorphisms in the *CLOCK* and other internal clock genes (and related haplotypes) are critically involved in the genetic susceptibility to obesity
- Systems biology analysis shows that the circadian system-related genes and the predicted related proteins are highly involved in the regulation of the transcriptional activity of RNA polymerase II probably as nuclear factors "per se" or through their activity on epigenetic marks, i.e., histone deacetylases, SIRTs, etc.
- Serotonin transporter is involved in the modulation of the obesity phenotype and shows a significant interaction with the master gene *CLOCK*. These findings may have therapeutic implications.
- The circadian system is linked to several master regulator of energy metabolism. One of the most important being the transcriptional coactivator, peroxisome proliferative activated receptor gamma coactivator 1 alpha, also known as PGC1a (*PPARGC1A*). Epigenetic regulation of its activity may be important in the disease process. *PPARGC,* and other nuclear factors, are critically involved in mitochondrial function which seems to be compromised in the MetS.

Acknowledgments This study was partially supported by grants PICT 2008-1521 and 2010-0441 (Agencia Nacional de Promoción Científica y Tecnológica), and UBACYT CM04 (Universidad de Buenos Aires). S.S. and C.J.P. belong to CONICET.

References

1. Reaven GM (1998) Insulin resistance may or may not play a role in blood pressure regulation. J Intern Med 244(4):359–360
2. Sookoian S, Gianotti TF, Schuman M, Pirola CJ (2009) Gene prioritization based on biological plausibility over genome wide association studies renders new loci associated with type 2 diabetes. Genet Med 11(5):338–343
3. Sookoian S, Pirola CJ (2011) Metabolic syndrome: from the genetics to the pathophysiology. Curr Hypertens Rep 13(2):149–157
4. Turek FW, Joshu C, Kohsaka A, Lin E, Ivanova G, McDearmon E et al (2005) Obesity and metabolic syndrome in circadian Clock mutant mice. Science 308(5724):1043–1045

5. Eckel-Mahan KL, Patel VR, Mohney RP, Vignola KS, Baldi P, Sassone-Corsi P (2012) Coordination of the transcriptome and metabolome by the circadian clock. Proc Natl Acad Sci U S A 109(14):5541–5546
6. Sookoian S, Gemma C, Gianotti TF, Burgueno A, Castano G, Pirola CJ (2008) Genetic variants of Clock transcription factor are associated with individual susceptibility to obesity. Am J Clin Nutr 87(6):1606–1615
7. Scott EM, Carter AM, Grant PJ (2008) Association between polymorphisms in the Clock gene, obesity and the metabolic syndrome in man. Int J Obes (Lond) 32(4):658–662
8. Garaulet M, Corbalan MD, Madrid JA, Morales E, Baraza JC, Lee YC et al (2010) CLOCK gene is implicated in weight reduction in obese patients participating in a dietary programme based on the Mediterranean diet. Int J Obes (Lond) 34(3):516–523
9. Garaulet M, Lee YC, Shen J, Parnell LD, Arnett DK, Tsai MY et al (2009) CLOCK genetic variation and metabolic syndrome risk: modulation by monounsaturated fatty acids. Am J Clin Nutr 90(6):1466–1475
10. Sookoian S, Castano G, Gemma C, Gianotti TF, Pirola CJ (2007) Common genetic variations in CLOCK transcription factor are associated with nonalcoholic fatty liver disease. World J Gastroenterol 13(31):4242–4248
11. Maury E, Ramsey KM, Bass J (2010) Circadian rhythms and metabolic syndrome: from experimental genetics to human disease. Circ Res 106(3):447–462
12. Stubblefield JJ, Terrien J, Green CB (2012) Nocturnin: at the crossroads of clocks and metabolism. Trends Endocrinol Metab 23(7):326–333 [Epub ahead of print]
13. Barbosa-Silva A, Fontaine JF, Donnard ER, Stussi F, Ortega JM, Andrade-Navarro MA (2011) PESCADOR, a web-based tool to assist text-mining of biointeractions extracted from PubMed queries. BMC Bioinformatics 12:435–444
14. Vidal M (2005) Interactome modeling. FEBS Lett 579(8):1834–1838
15. Chen J, Bardes EE, Aronow BJ, Jegga AG (2009) ToppGene Suite for gene list enrichment analysis and candidate gene prioritization. Nucleic Acids Res 37(Web Server issue): W305–W311
16. Sookoian S, Gemma C, Pirola CJ (2010) Influence of hepatocyte nuclear factor 4alpha (HNF4alpha) gene variants on the risk of type 2 diabetes: a meta-analysis in 49,577 individuals. Mol Genet Metab 99(1):80–89
17. Sookoian S, Gianotti TF, Gemma C, Burgueno AL, Pirola CJ (2010) Role of genetic variation in insulin-like growth factor 1 receptor on insulin resistance and arterial hypertension. J Hypertens 28(6):1194–1202
18. Sookoian S, Gemma C, Garcia SI, Gianotti TF, Dieuzeide G, Roussos A et al (2007) Short allele of serotonin transporter gene promoter is a risk factor for obesity in adolescents SOOKOIAN2007A. Obesity (Silver Spring) 15(2):271–276
19. Sookoian S, Gianotti TF, Gemma C, Burgueno A, Pirola CJ (2008) Contribution of the functional 5-HTTLPR variant of the slc6a4 gene to obesity risk in male adults SOOKOIAN2008. Obesity (Silver Spring) 16(2):488–491
20. Burgueno A, Gemma C, Gianotti TF, Sookoian S, Pirola CJ (2010) Increased levels of resistin in rotating shift workers: a potential mediator of cardiovascular risk associated with circadian misalignment. Atherosclerosis 210(2):625–629
21. Sookoian S, Gemma C, Fernandez GT, Burgueno A, Alvarez A, Gonzalez CD et al (2007) Effects of rotating shift work on biomarkers of metabolic syndrome and inflammation SOOKOIAN2007. J Intern Med 261(3):285–292
22. Sookoian S, Pirola CJ (2009) Shift work and subclinical atherosclerosis: recommendations for fatty liver disease detection. Atherosclerosis 207(2):346–347
23. Warde-Farley D, Donaldson SL, Comes O, Zuberi K, Badrawi R, Chao P et al (2010) The GeneMANIA prediction server: biological network integration for gene prioritization and predicting gene function. Nucleic Acids Res 38(Web Server issue):W214–W220
24. Esterbauer H, Oberkofler H, Krempler F, Patsch W (1999) Human peroxisome proliferator activated receptor gamma coactivator 1 (PPARGC1) gene: cDNA sequence, genomic organization, chromosomal localization, and tissue expression. Genomics 62(1):98–102

25. Finck BN, Kelly DP (2006) PGC-1 coactivators: inducible regulators of energy metabolism in health and disease. J Clin Invest 116(3):615–622
26. Sookoian S, Garcia SI, Porto PI, Dieuzeide G, Gonzalez CD, Pirola CJ (2005) Peroxisome proliferator-activated receptor gamma and its coactivator-1 alpha may be associated with features of the metabolic syndrome in adolescents. J Mol Endocrinol 35(2):373–380
27. Sookoian S, Rosselli MS, Gemma C, Burgueno AL, Fernandez GT, Castano GO et al (2010) Epigenetic regulation of insulin resistance in nonalcoholic fatty liver disease: impact of liver methylation of the peroxisome proliferator-activated receptor gamma coactivator 1alpha promoter. Hepatology 52(6):1992–2000
28. Li S, Zhao JH, Luan J, Luben RN, Rodwell SA, Khaw KT et al (2010) Cumulative effects and predictive value of common obesity-susceptibility variants identified by genome-wide association studies. Am J Clin Nutr 91(1):184–190
29. Tomaszewski M, Charchar FJ, Lacka B, Pesonen U, Wang WY, Zukowska-Szczechowska E et al (2004) Epistatic interaction between beta2-adrenergic receptor and neuropeptide Y genes influences LDL-cholesterol in hypertension. Hypertension 44(5):689–694
30. Baessler A, Fischer M, Mayer B, Koehler M, Wiedmann S, Stark K et al (2007) Epistatic interaction between haplotypes of the ghrelin ligand and receptor genes influence susceptibility to myocardial infarction and coronary artery disease. Hum Mol Genet 16(8):887–899
31. Isaacs A, Aulchenko YS, Hofman A, Sijbrands EJ, Sayed-Tabatabaei FA, Klungel OH et al (2007) Epistatic effect of cholesteryl ester transfer protein and hepatic lipase on serum high-density lipoprotein cholesterol levels. J Clin Endocrinol Metab 92(7):2680–2687
32. Brisson D, St-Pierre J, Santure M, Hudson TJ, Despres JP, Vohl MC et al (2007) Genetic epistasis in the VLDL catabolic pathway is associated with deleterious variations on triglyceridemia in obese subjects. Int J Obes (Lond) 31(8):1325–1333
33. Sookoian S, Gianotti TF, Burgueno A, Pirola CJ (2010) Gene–gene interaction between serotonin transporter (SLC6A4) and clock modulates the risk of metabolic syndrome in rotating shiftworkers. Chronobiol Int 27(6):1202–1218
34. Ling C, Del GS, Lupi R, Ronn T, Granhall C, Luthman H et al (2008) Epigenetic regulation of PPARGC1A in human type 2 diabetic islets and effect on insulin secretion. Diabetologia 51(4):615–622
35. Gemma C, Sookoian S, Alvarinas J, Garcia SI, Quintana L, Kanevsky D et al (2009) Maternal pregestational BMI is associated with methylation of the PPARGC1A promoter in newborns. Obesity (Silver Spring) 17(5):1032–1039
36. Gemma C, Sookoian S, Dieuzeide G, Garcia SI, Gianotti TF, Gonzalez CD et al (2010) Methylation of TFAM gene promoter in peripheral white blood cells is associated with insulin resistance in adolescents. Mol Genet Metab 100(1):83–87
37. Gemma C, Sookoian S, Alvarinas J, Garcia SI, Quintana L, Kanevsky D et al (2006) Mitochondrial DNA Depletion in Small- and Large-for-Gestational-Age Newborns. Obesity (Silver Spring) 14(12):2193–2199
38. Gianotti TF, Sookoian S, Dieuzeide G, Garcia SI, Gemma C, Gonzalez CD et al (2008) A Decreased Mitochondrial DNA Content Is Related to Insulin Resistance in Adolescents. Obesity 16(7):1591–1595

List of Bioinformatic Resources

PESCADOR: available at: http://cbdm.mdc-berlin.de/tools/pescador/
ToppGene Suite: available at: http://toppgene.cchmc.org
GeneMANIA: available at: http://www.genemania.org
KEGG: Kyoto Encyclopedia of Genes and Genomes: available at: http://www.genome.jp/kegg

Index

M. Garaulet and J.M. Ordovás (eds.), *Chronobiology and Obesity*,
DOI 10.1007/978-1-4614-5082-5, © Springer Science+Business Media New York 2013

Printed by Publishers' Graphics LLC
AMZ20121216.19.24.149